A SHORT INTRODUCTION  TO CLIMATE CHANGE

A Short Introduction to Climate Change provides a clear, balanced and well-documented account of one of the most important issues of our time. It covers developments in climate science over the past 250 years, compares climates over geologic time and shows that recent climate change is more than the result of natural variability. It explains the difference between weather and climate by examining changes in temperature, rainfall, Arctic ice and ocean currents. It also considers the consequences of our use of fossil fuels and discusses some of the ways to reduce further global warming. In this book, Tony Eggleton avoids the use of scientific jargon to provide a reader-friendly explanation of the science of climate change.

Concise but comprehensive, and richly illustrated with a wealth of full-colour figures and photographs, *A Short Introduction to Climate Change* is essential reading for anyone who has an interest in climate science and in the future of our planet.

Tony Eggleton is an Emeritus Professor of the Australian National Univer   in Science from the
Universi   'Philosophy
at th   ch into
min   by the

A SHORT INTRODUCTION TO

CLIMATE CHANGE

TONY EGGLETON

CAMBRIDGE
UNIVERSITY PRESS

CAMBRIDGE UNIVERSITY PRESS
Cambridge, New York, Melbourne, Madrid, Cape Town,
Singapore, São Paulo, Delhi, Mexico City

Cambridge University Press
477 Williamstown Road, Port Melbourne, VIC 3207, Australia

Published in the United States of America by Cambridge University Press, New York

www.cambridge.org
Information on this title: www.cambridge.org/9781107618763

First published 2013

Cover and text design by Denise Lane, Sardine Design
Copy-editing by Renée Otmar
Typeset by Newgen Publishing and Data Services
Printed in China by C & C Offset Printing Co. Ltd

A catalogue record for this publication is available from the British Library

*A Cataloguing-in-Publication entry is available from the catalogue
of the National Library of Australia at* www.nla.gov.au

ISBN 978-1-107-61876-3 Paperback

CONTENTS

PREFACE

I had a most satisfying career, able to do what I really wanted, paid to teach and do scientific research and free to explore where my curiosity led me. That is the working life of many a scientist, and I was certainly one of the lucky ones. Now, I had never taken much notice of climate or the weather, at least no more than anyone whose day out is spoilt by rain. True, I did spend six years in Wisconsin, in the United States, where much of the evidence for the Great Ice Ages lies across the landscape, and in another way the last 30 years of my research have indeed been about weather. Not the coming and going of storms and droughts, but the effect the weather has upon rocks. You see, I am a geologist, and I was led by curiosity, coincidence and colleagues into the study of the way in which rocks become soil – mineral weathering. I looked deep into the heart of clays, photographing their relationship to the harder minerals out of which they grew with what was then the most amazing microscope in the world. On a wall in my office are images of minerals magnified to the point at which their constituent atoms are visible. When I retired, there were new opportunities to explore, from playing with grandchildren to travelling the grand landscape of Australia. And so it turned out. But after tidying up the loose ends of 40 years' scientific research, in the longest drought I have experienced and amidst a political storm over the evidence for global warming and a proposed Emissions Trading Scheme, I received a challenging message by email. In this book you will read about where that led.

Two years of reading and writing have taken me away from my wife, Glen, for countless hours – hours when we might have been doing more of those other things. For letting me work without complaint, for supporting me and for reading and gently improving what I wrote, my heartfelt thanks.

Our daughters, Rachel and Cate, added their wisdom and encouragement as each chapter appeared, and again as they were shuffled into their final sequence. Paul Coromandel also read the draft manuscript; his comments were instrumental in improving the text. The figures would still be pencil sketches if Geoff England had not shown me the finer points of computer draughting. My sister, Hilary, and cousins, John and Barbara and her husband Anton, also suggested improvements in the name of clarity. My family was of paramount importance, not just for suggesting ideas and changes, but also for pushing me away from my customary lecturing style into what I hope is an engaging writing style. I am extraordinarily grateful to them all.

Colleagues at the Australian National University also advised me and informed the development of the book. First was Andrew Glikson, whose passion for the subject was infectious and whose fount of knowledge and resources made starting the job so much easier. Brad Opdyke, John Fitz-Gerald, Patrick De Deckker and Paul Tregoning listened to my questions and always had the answers.

There are a number of scientists whom I have never met but who generously and promptly replied to my emailed pleas for help. To Paul Halloran, Caroline Ummenhofer, Donna Roberts, Judith Lean, David Post, Claudine Chen, Sophie Dove and Ray Langenfelds, thank you for clarifying many issues.

Then, the first draft was completed and it needed peer review. Professor Will Steffen from my university's Climate Change Institute most generously agreed to read and critique what I had written. My thanks go to him for finding the time to read and correct that draft, for his positive response and for encouraging me to find a publisher. That took rather some time, and when I was discouraged it was Dr Barry Fordham who not only gave me the kick that sent me to Cambridge University Press, but then also read every word of the second draft and gave me succinct and sage advice on almost every paragraph. For the generous gifts of your time and wisdom, Barry, I am forever indebted. The text was greatly improved by the careful and detailed comments, corrections and advice given to me by Professor Martin Williams, to whom I am most grateful. My utmost thanks go to Professor Tom Wigley. At short notice he agreed to review the manuscript, and his vast experience and knowledge of the science of climate change were of immense help to me in the final stages of editing.

For photographs, my thanks to Hilary, John and Andrew Brooke, to Roger Hambly, Paul Halloran, Donna Roberts, Daniel Steiner, Ove Hoegh-Guldberg, James Morrow and Giselle Coromandel. I am grateful for the support of the Australian National University, through Professor Andrew Roberts, Director of the Research School of Earth Sciences, for a subsidy to help print this work in colour.

Finally, to my oldest friend, Sandy Facy, thanks. You got me started on the book with a question – and had to wait years for the answer.

1

THE SPIRIT OF ENQUIRY

The inquiry of truth ... is the sovereign good of human nature.

Francis Bacon

Just before Christmas 2009, an old friend and I were discussing climate change. Because I am a scientist, he asked my opinion about an internet site claiming that global warming is a fallacy and that carbon dioxide is good for the planet. As a geologist I knew about climate changes of the distant past, such as the warm world when dinosaurs roamed and the more recent ice ages, but I could not answer some of the questions he asked, and that prompted me to look into climate science more deeply. My reply to him is this book.

Looking at the newspaper articles and websites on the topic, I found a remarkably wide range of views, running the gamut from 'global warming is real, burning coal and oil caused it, we are doomed' to 'we need to address the reality that increasing amounts of greenhouse gases are causing climate change' and on to 'increased levels of carbon dioxide will grow bigger crops and stop the next ice age'.

The predictions and politicking surrounding the subject confused me, and I needed to find out what is actually known about the subject. Why was it that climate scientists were so concerned about the future? How were they able to predict the course of climate change so persuasively that many of the world's governments were prompted to at least think about the problem, to establish an intergovernmental panel on the subject, to gather at Copenhagen in 2009 to discuss climate change and for many governments to pass legislation designed to limit carbon dioxide emissions?

In the years leading up to that initial discussion with my friend, climate change had become a political issue. Just as everyone had a right to their own political opinions, everyone seemed to discover a right to their own opinion on climate change. Very quickly, two opposing camps were established: in one camp were those who accepted climate change on the basis of the science, and in the other were those who denied the validity of the scientific evidence. But many people were in neither camp. They heard one side, they heard the other, and like my friend they were unsure about whom to trust.

Many of those who deny the science of climate change call themselves 'sceptics'. But all good scientists are sceptics. A scientist's first scepticism is usually aimed at his or her own evidence; unusual results and remarkable conclusions need to be challenged immediately, long before they are made public. Scepticism should involve argument – argument drawing on *all* the relevant evidence. When an argument uses only some parts of the evidence – parts that appear to support the case – the argument becomes adversarial, not sceptical. It is not a defence lawyer's job to present evidence that will convict the defendant. Rather, the defence selects only evidence that it hopes will deny the prosecution. True sceptics work differently; they query and challenge *all* the evidence.

The website that got me started is run by Leon Ashby, president of a political party called the Climate Sceptics, and it quickly became clear to me that this party's approach is adversarial, not sceptical. As an example, one of its claims, which is repeated by several others, including Professor Ian Plimer, author of the book *Heaven and Earth. Global Warming: The Missing Science* (Connor Court, 2009, from here on referred to as '*heaven+earth*'), and Christopher Monckton (Third Viscount Monckton of Brenchly, often referred to as 'Lord Monckton'), to name two who have achieved prominence on the subject, is that global warming ceased in 1998.

In 1990, the United Nation's Intergovernmental Panel on Climate Change (IPCC) predicted that global temperature would rise about 0.4°C by 2010. The sceptics argue, however, that from 1998 to 2008 global temperature fell by 0.2°C; all the IPCC's predictions are rubbish and there is no global warming. The evidence the Climate Sceptics present for this conclusion is shown in Figure 1.1a.

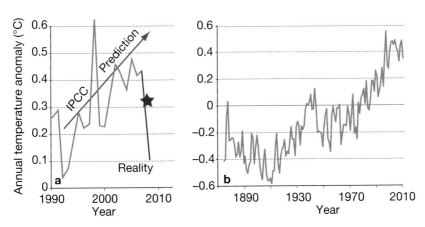

Figure 1.1 a) Global cooling since 1998 copied (in essence) from the Climate Sceptics presentation. I have added the star, the 'reality' for 2008. b) Annual global temperature anomaly from 1880 to 2010 drawn from data published by the Hadley Climatic Research Unit at the University of East Anglia and the Hadley Centre for Climate Prediction and Research, United Kingdom.

The IPCC projection was based on temperature measurements over more than a century, shown in Figure 1.1b. The Climate Sceptics' argument in effect 'cherry-picked' only the temperatures for the period around 1998, the hottest year on record. Compared to that peak, world temperatures since then have, indeed, been lower. The Climate Sceptics base their argument on two non-scientific interpretations of the data: one ignores data that do not support its hypothesis, in this case all data from before 1990, and the other mistakes the natural variations from year to year as indicative of variations in climate. Ten years is too short to conclude *anything* about climate change, because it is the weather conditions over a long period – 30 to 50 years – that defines the climate.

One hundred and thirty years of observations does provide a reliable climate history. Though there are some irregularities, the graph in Figure 1.1b indicates that the global temperature is rising and it cannot be ignored. Climate science explains the temperature rise as being largely the result of burning coal and oil for the past 200 years, thereby putting the greenhouse gas carbon dioxide into the atmosphere. 'Not so,' say the sceptics. 'The Earth has experienced many changes in temperature over geological time without any contribution by people.' Well, that is true, and you will read in this book about the climate changes of the geological past and why they occurred. Other views are presented, also; *El Niño* events, variations in the Sun's output and cosmic or galactic factors have all been proposed as more reasonable explanations for global warming.

THE HOCKEY STICK

The most famous evidence for global warming followed from a 1998 paper published in the major scientific journal *Nature* by Michael Mann, then a post-doctoral fellow at the University of Massachusetts, with professors Raymond Bradley and Malcolm Hughes. These authors presented a graph of global temperatures over the past six centuries, and a later refinement presented in 2001 (Figure 1.2) has become known as the 'Hockey Stick'.[1] By combining thermometer measurements taken since 1850 with evidence from earlier times, Mann and his colleagues concluded that the past 100 years have been a century of remarkable warming. On the basis of documented determinations of temperature, they showed that the Earth's climate, although fluctuating, had been cooling overall since mediaeval times, 1000 years ago. For 900 years the global temperature fell slowly, though with small ups and downs, by about 0.3°C, but then there was a change. During the 20th century, the global temperature rose by 0.7°C. As is the way of science, the Hockey Stick graph had its critics, and the scientists responded by collecting more data and refining their conclusions. When further work is shown to be needed, scientists do it, for if there are any problems with their work they want to fix it. Changing one's conclusions does not discredit science, rather it enhances and strengthens it. We will return to the Hockey Stick graph in Chapter 9; in the years since it was first published its shape, its conclusions and its significance for our future have all been confirmed many times over.

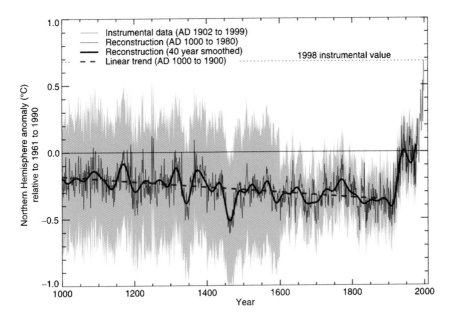

Figure 1.2 The Mann Hockey Stick temperature graph as presented in Figure 2.20 in the *Intergovernmental Panel on Climate Change Third Assessment Report* (2001). The grey shaded area represents the likely error of each data point, while the dashed red line shows the overall trend from the year 1000 to 1900.

In this book you will see just how detailed is the understanding of the world's climate, and you will learn the degree of reliability that can be attributed to the science. Figure 1.2 is such an example. The measured or estimated temperatures join to form the zigzag blue line, with thermometer measurements in red. The grey shading shows the scientists' judgement about the reliability of each measurement. Their best conclusion about the most likely value at any time is the wavy black line. And the dashed red line shows the trend up to 1900. There are ups and downs in the global temperature graph; each is a response to known factors explained in Chapter 3.

From my own point of view as a scientist, perhaps the most disappointing thing about those who criticise the climatologists, rather than their results or conclusions, is the repeated assertions that the scientists 'fudge' their data to satisfy those who fund their research. Here is Monckton's view, expressed shortly after newspaper reports of allegedly improper email correspondence between scientists at the Hadley Climatic Research Unit at the University of East Anglia, an issue which became known as 'climategate'.

> Many of these organizations are deeply implicated in the Climategate scandal. The emails between them have demonstrated a systematic, self-serving, ruthless readiness to invent, fabricate, distort, alter, suppress, hide, conceal or even destroy scientific data for the sake of reaching the answer they want.[2]

Following the allegation, three independent reviews were promptly conducted of the Research Centre's activities, and all three concluded that there was no evidence of malpractice, Director Phil Jones had no case to answer and those who criticised the scientists had been selective and uncharitable. Lord Monckton is influential in the public domain because newspapers print his views, but his comments in relation to 'climategate' are unsubstantiated.

Scientists are human, they have failings, and there have been examples of scientific fraud, some well documented. However, by far the majority of climate science is done by groups, all of whom must be confident of the reliability of the work of their colleagues. It is difficult to imagine that all have conspired to deceive both their reviewers of their papers and the scientific scrutiny that follows publication.

I am a geologist with a particular interest in near-surface weathered rocks, and I have not published findings on climate science. However, I have a comprehensive understanding of the scientific process, how scientific results are presented and the scrutiny they undergo before they are published. Usually, a scientific paper is presented in four parts:

1) A statement of the purpose of the study and how it fits into the body of knowledge of that scientific field

2) How the study was conducted: there must be enough detail for the study to be replicated by others if they so choose

3) What the study discovered, including all the results that have a bearing on the purpose, not just those that fit its hypothesis

4) What the authors think the results mean.

Every such manuscript submitted to a scientific journal undergoes peer review. Typically, there are at least two reviews by experts in the field who are selected by the journal editor and who are known not to be colleagues of the authors. Quite commonly a manuscript is subjected to more than two reviews, as well as the one by the editors of the journal. Reviewers tend to be particularly scrupulous about requiring appropriate and adequate descriptions of experiments, reference to relevant work by others and that conclusions are based on data presented within the manuscript.

Scientists tend not to report their results without a string of 'ifs' and 'buts', and this could make it appear that the research is doubtful or that the conclusions are suspect. However, it is not in the nature of science to be dogmatic and assertive, but rather to present the evidence, to weigh and consider it, point out its limitations and report conclusions on the basis of what has been learned.

It is important to remember that scientists are people who possess the range of human emotions like all of us. They do not like being wrong, they do like being right, and they tend always to be ready to correct others. 'Told you so' is just as popular among scientists as within families! When a scientific paper is published with a contentious idea a flood of counter claims and contrary opinions will be quickly offered, challenging the methods, the results and the ideas. No new concept survives without intense scrutiny, both before and after it has been published. And when an idea is criticised, there is never any suggestion of scientists closing ranks to protect their colleagues, for the attacks mostly come from fellow scientists. When, in 1989, an experiment purporting to show nuclear fusion had been achieved at room temperature, it was scientists who pointed out the flaws in the work. When in 2011 neutrinos travelling faster than light were reported, again it was scientists who questioned the results, and in this case it was the researchers themselves who found the mistake.

It is said that you can never prove a hypothesis right, but you can always prove it wrong. But that is only true to an extent. Some hypotheses are of the type that involves a choice between only two possibilities. If one is shown *not* to be true, the other must be so. Such an hypothesis stirred the geological community from 1915 to at least the 1970s. In 1915, Alfred Wegener published a book on the subject of drifting continents. He had noticed the jigsaw puzzle fit between the shapes of South America and Africa, and between North America and Europe, and had seen how the rocks and the fossils matched each other across the Atlantic Ocean. He concluded the two huge landmasses had once been united (see Figure 1.3) and had since drifted apart.

Figure 1.3 Pangaea, the super continent that broke up 200 million years ago, allowing the continents as we know them to drift apart.

Wegener was variously ridiculed or accepted, but mostly derided. The geological world polarised into two camps: those who thought the hypothesis of continental drift was probably right, and those who, unable to conceive a mechanism for this, refused to consider it. 'Drifters' tended to live in the southern continents, where much of the evidence in the rocks was to be found. 'Fixers' were mostly in the United States and Europe, though by 1960 Cambridge and Edinburgh universities certainly had drifters among their geologists. I was studying at the University of Wisconsin in Madison, United States in 1963 when a new geologist was appointed to the staff, a young man from Edinburgh in the United Kingdom. In his inaugural seminar he talked about continental drift and the recent evidence in favour of it. As I left the lecture theatre I walked behind two of the department's senior professors. Their voices were troubled and they were shaking their heads. 'What have we appointed!'

I overheard one say. So strong was the conviction of the fixers that new evidence at first profoundly disturbed them. But as the years passed and the evidence mounted, so their understanding changed.

Continental drift, now called 'plate tectonics', was an 'either/or' theory; either the continents drift or they do not. If instead of 'the continents are drifting' your theory is that 'they are fixed', a single experiment with a global positioning system (GPS) instrument can disprove the 'fixed' and logically prove the 'moving' theory. The continents drift at about 5 to 10 centimetres a year, the same speed as our fingernails grow. The theory of continental drift *has* been proven.

How does the theory of global warming stand in this regard? The opposite theory would be 'the average annual global temperature is not increasing'. Is there evidence that might disprove this theory? You could ask 'how much change and over how long a period would be enough to prove an increase?' For continental drift, the change is about half a kilometre in 100 years. Would 0.5°C in 100 years be enough to disprove the 'no warming' hypothesis?

This book is written for those who hear the arguments for and against climate change and are not sure about them. It is written for sceptics, a term derived from the Greek word for 'enquiry'. This book is written for enquirers. If you want to know what the science is about, if you want to understand climate and climate change, here you will find the issues presented as fairly as I can. I will not be selecting the reports that support (or deny) any particular view of the subject, but I will be selecting reports that have the authenticity that comes from peer review and then public exposure. By 'authenticity' I do not mean that I think the conclusions of the work are 'right'; only that I think the work is good science. The data, the observations and the measurements all have to be good. Good data can last for generations; their interpretation can be anything from brilliant to wrong. I wonder how the theory of climate change caused by the burning of fossil fuels will be viewed in 100 years from now. Interpretations evolve, change and sometimes settle into accepted fact: the Sun *is* at the centre of the solar system, the continents *have* drifted and smoking *does* damage the lungs.

FURTHER READING

For this and all subsequent chapters, the Fourth Assessment Reports of the Intergovernmental Panel on Climate Change (IPCC, 2007) provides detailed summaries and analyses of the science of climate change.

Start with *Climate Change 2007: Synthesis Report*, then for further detail see:

Cleugh H, Stafford Smith M, Battaglia M & Graham P (eds) (2011) *Climate Change: Science and Solutions for Australia*. CSIRO Publishing, free download available at www.csiro.au/Climate-Change-Book

Climate Change 2007 – The Physical Science Basis Contribution of Working Group I to the Fourth Assessment Report of the IPCC (ISBN 978 0521 88009–1 hardback; 978 0521 70596–7 paperback)

Climate Change 2007 – Impacts, Adaptation and Vulnerability Contribution of Working Group II to the Fourth Assessment Report of the IPCC (978 0521 88010–7 hardback; 978 0521 70597–4 paperback)

Climate Change 2007 – Mitigation of Climate Change Contribution of Working Group III to the Fourth Assessment Report of the IPCC (978 0521 88011–4 hardback; 978 0521 70598–1 paperback)

Pittock, AB (2009) *Climate Change: The Science, Impacts and Solutions*. CSIRO Publishing.

GLOBAL WARMING

To set budding more
And still more later flowers for the bees
Until they think warm days will never cease.

John Keats

We could now be entering a new stage in the history of the Earth. From the dawn of time our planet's climate has been changed by the life forms it spawned. There was no oxygen in the atmosphere of the early Earth. When microscopic plants evolved in the primordial oceans they generated oxygen, and from then the atmosphere was never the same again. Later, in the warm days, as land plants evolved they drew down abundant carbon dioxide, eventually causing an ice age. How ironic that by burning their long-buried remains we are returning their carbon to the air and in turn we could be changing the world.

There is one vital difference between the climate changes of geologic time and that of today: speed. As we emerged from the last ice age, the average annual global temperature rose by an average of 1°C each 1000 years, then it started to cool again quite slowly, at about two-tenths of a degree each 1000 years. We will see that now it is changing 25 times as quickly *in the opposite direction*.

Historical records show that the European climate was warmer in mediaeval times than it was during the so-called Little Ice Age, from 1500 to 1850. This cooling, and a slight fall in global temperature after the Second World War, led to a handful of scientific studies in the 1970s suggesting that the Earth was headed toward the next ice age, some 20 000 years ahead. This recognition of climate cycles was picked up by the media and inflated into dire predictions of imminent freezing; stories that of course proved false but that are used by some to discredit the reports and projections of climate scientists of the 21st century. In fact, most climate scientists of the 1970s were writing of warming trends, not of an imminent ice age.

The concept of climate change is certainly not new, but by the late 20th century it had moved from being of interest only to historians and palaeontologists to being a topic of almost daily conversation. Global warming is now at the heart of a political and social debate about human involvement in climate change. It has polarised opinion, divided friends, created arguments and possibly caused changes in governments.

There seem to me to be three main questions that need to be asked:

1) Is the climate is changing? We will look at the evidence in this chapter and in Chapters 5, 6 and 7.

2) What can change the climate? Chapters 3 and 4 will cover this topic.

3) How has climate changed in the past? In Chapters 8 and 9 we will view climate through geologic time.

Change is only meaningful by comparison with some previous condition. A grandmother will exclaim to a grandchild she has not seen for a year: 'My, how you have grown.' A stranger, upon being introduced to the same child, would not think of making this remark. Similarly, in order to assess climate change we need to have knowledge of the climates of the past. But of course climate differs from place to place around the world. Southern Australia is never hit by cyclones, northern Australia is not drought-prone, Tasmania gets most of its rain in winter and Cape York in Queensland gets its rain in summer. Climate change might result in more rain occurring in one place, less rain in another and no change in a third. To answer question 1), climate comparisons have to be made over time for many specific places, and that investigation will also answer question 2).

BLOSSOMS, BUTTERFLIES AND BIRDS

Nature herself has provided the clues that reveal that a change in the climate has been occurring during the 20th century. In climates in which life pauses for the cold of winter, plants and animals respond to the warmth of spring. Trees bud, frogs spawn, birds mate and caterpillars hatch from their dormant eggs. And if spring comes early, so does the re-awakening of life.

Cherry blossom time is one of Japan's most popular tourist attractions, and the blossom has long been an important artistic element in that country's culture. So popular are local cherry blossom festivals that the Japan Meteorological Agency tracks the date of the opening of the first flowers as it moves northward during spring. Records of the first blossoms date back to 1400. Cherry blossom time now begins a week sooner than it did in 1950.[1]

The Burgundy region of France is arguably the world's best-known wine district. Harvest time has been at the heart of religious and secular autumn festivals throughout the region's history and, not surprisingly, records of the date of the grape harvest extend as far back as the 16th century. In 1800, the harvest in Burgundy was typically held in the first week of October. By 1950 harvest took place in the last week of September, and by 2000 it was almost another week earlier.[2] The 2011 harvest was the earliest on record, beginning at the end of August.[3] Grapes grown in the region are now ripening more than a month earlier than they did 200 years ago.

Field naturalists' clubs abound in Europe; many have been in existence since the 19th century. One consequence of these clubs' activities is that there is now a long record of springtime events, such as the sound of the cuckoo's first call for the season, the first appearance of a butterfly or the first opening of the crocus flower. Migrating sand martins now arrive in England 20 days earlier than they did 50 years ago. Almost all British butterflies now emerge one or two weeks earlier than they did 100 years ago.[4] The green-veined white butterfly (Figure 2.1) has been studied in some detail. In 1975 this butterfly typically emerged from its chrysalis in early June, then over the ensuing years it appeared earlier and earlier, until by 1998 it was emerging by the end of the first week in May.[5] British data spanning the past 250 years on the first flowering of 450 plant species have been combined by Amano and colleagues.[6] They found the average date of first flowering to be almost two weeks earlier than 150 years ago.

Figure 2.1 Green-veined white butterfly. This species is now appearing three weeks earlier in the spring than it did only 25 years ago.

Insect species are confined to places where the climate is right for them. In the case of butterflies, this may be because caterpillar-food plants thrive in that climate or the caterpillar itself survives best there. Whatever the butterfly's reason, one study has shown that of 57 European butterflies, two-thirds have moved their range northward over the past century, some by as much as 200 kilometres.[7] Such a change implies a move in order to remain within the insect's preferred climate range, which in turn implies that the climate has changed, and in this case because it is getting warmer.

Other examples of changes in the behaviour of plants and animals abound. In Canada the aspen poplar is flowering almost a month earlier than it did in 1900. In a study spanning almost 50 years, the father and son team of R.S.R. Fitter and Professor Alastair Fitter compared the first flowering date of 385 species of plants, mostly from one rural area in England, Oxfordshire. They found that the average first flowering date during the decade 1991–2000 was 4.5 days earlier than it had been in the period 1954–90. An extreme example of change is the perennial plant white deadnettle, which occasionally used to flower in winter, but now routinely does so. Another is the ivy-leaved toadflax, which now flowers in mid-March whereas 50 years ago it did not flower until mid-April.[8]

Another indicator of warming comes from the Friesland province of The Netherlands. Since 1909, a skating race called the *Elfstedentocht* (eleven cities' tour) has been held along the frozen canals during winter. The last race, held in 1997, attracted about 16 000 participants. The race can only be run when the ice is thick enough (15 centimentres) over the entire course, and between 1909 and 1959 the winters were cold enough for 10 races to be held. Over the following 50 years the race was only run four times.[9]

In Melbourne, Victoria, Australia's Common Brown butterfly (*Heteronympha merope*) now emerges from its winter chrysalis about 10 days earlier than it did in 1950. Experiments with the growth of this insect's caterpillars at different temperatures confirm that increasing the average temperature accelerates the caterpillar's life cycle.[10]

Insect activity and the budding and flowering of plants in spring are influenced by the temperature. If many different kinds of insect are becoming active earlier in the spring than they have for centuries and if plants are flowering earlier, the only reasonable explanation is that winter and spring are getting warmer. Some of the early appearances might be caused by the 'urban heat-island effect'. The green-veined white is a common garden butterfly, and specimens over-wintering in an urban garden might be kept warm by the heat from the surrounding houses and so develop earlier than their country cousins. This argument certainly cannot hold for the Burgundy grape harvest, for migrating birds, for freezing canals, or for the many varieties of flowering plants in Oxfordshire assessed by the Fitters. Plants and animals are telling us that their world is warming, a story that is independent of scientists' approach to measuring phenomena. Can we measure whether this is indeed the case?

TEMPERATURES SINCE 1880

The temperature on a given day at any place depends on many things. First is the difference between day and night, then how warm the sun is when it is out; whether the sun is high in the sky and hot in summer or lower and cooler in winter. Clouds shield the sun's heat by day and keep the ground warmer at night. The wind brings air from somewhere else, which might have a different temperature. Fires or dust storms over the previous week might have left particles suspended in the air, and these could absorb some of the sun's warmth. Determining a meaningful value for average temperature demands that measurements be made at the same place regularly and over a long period of time.

Temperatures have been measured with a liquid-in-glass instrument since about 1700, a time when a variety of different liquids and different scales were proposed. Not until the 19th century did the use of mercury or alcohol in glass thermometers graduated by degrees centigrade come into common usage.[11] As far as reliable and reproducible measurements of air temperature are concerned, there seem to be few records before about 1880 that are acceptable. The Australian Bureau of Meteorology has established about 100 Reference Climate Stations located away from urban areas, mostly with records held from 1910, for high-quality, long-term climate monitoring.

The direct daily measurement of temperatures that have been recorded at many weather stations around the world provide vital data about global temperature change. From the daily average temperature (maximum + minimum divided by two) a monthly average is calculated, and from that an annual average. But there is no sense in directly comparing the annual average temperature in tropical Darwin with the temperature in sub-Arctic Oslo. What is sensible is to compare local annual average temperatures as the years go by. This brings all the results to the same basis. The records for the city of Wagga Wagga in New South Wales can be used as an example. From 1910 to 1966, the temperature average for the 56 years was 15°C. That average could be used as the baseline to compare any changes after 1966. Over the following 43 years, the average temperature in Wagga Wagga, though still showing the variability introduced by weather, steadily rose. The year 2009 was 2.3°C warmer than the baseline average; the difference from the selected baseline is called the 'temperature anomaly'. A graph of all these variations is shown in Figure 2.2a.

When it comes to collating the measurements from all the world's measuring stations, it is the temperature anomaly for each station that is used rather than the actual temperature. Doing the same thing for Alice Springs in the Northern Territory yields a 2009 temperature anomaly

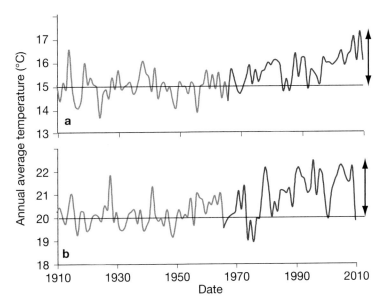

Figure 2.2 a) Temperature variation at Wagga Wagga, NSW. The left-hand side of the graph (1910–65) has been chosen as the baseline (average) against which the later part is compared. The temperature anomaly in 2009 of 2.3°C is shown as the difference above the baseline (arrow). b) Temperature variation for Alice Springs, NT. The temperature anomaly in 2009 was 2°C. Source: Australian Bureau of Meteorology.

of 2°C (Figure 2.2b). In this way all the results can be brought to the same level.

Expanding these observations to all of Australia provides national temperature trends since 1910. The black line in Figure 2.3 (p. 18) is the annual average as determined by the Australian Bureau of Meteorology; the coloured bars are the annual average differences from that average, in this case using the period 1961–90 as the baseline. Just as the Wagga Wagga and Alice Springs temperatures showed in Figure 2.2, there has been appreciable warming since 1960. The Bureau reports 2009 as being the second-warmest year on record nationally, after 2005.

It is noticeable how big the differences are between years, something that was noticeable also in the Wagga Wagga and Alice Springs graphs. Generally, the reasons for the variations are straightforward; for example, 2000 was a very strong *La Niña* year, and that had a cooling effect; 1998 was a strong *El Niño* and 2011 was another strong *La Niña* year. The impacts of *El Niño* and *La Niña* on the weather are discussed in Chapter 3.

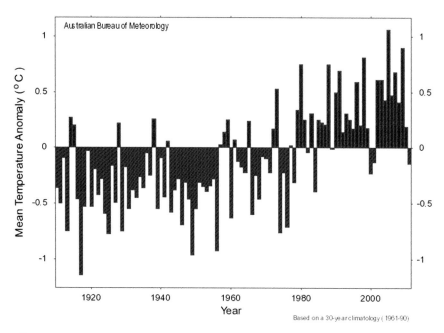

Figure 2.3 Australian mean temperature anomalies (difference from the 1961–90 average). Source: Australian Bureau of Meteorology, product of the National Climate Centre.

GLOBAL TEMPERATURE MEASUREMENT

Worldwide, there are enough reliable records to establish a global estimate, year by year. Every recording station has an established average over a defined reference period, as was done above for Wagga Wagga and Alice Springs, and the difference from that average yields the anomaly for each year. Once all of these station records are on the same scale, they are averaged to give the global temperature anomaly for that year.

In about 1980, a team led by James Hansen and funded by the United States National Aeronautics and Space Administration (NASA), took some previous analyses of northern hemisphere data and added what was available from the sparser southern hemisphere records. At the same time, a second team in England, led by Phil Jones and Tom Wigley, began an independent analysis of global temperature data. The two groups found the same thing: temperatures were rising and 1981 was the warmest year in 100 years.

Hansen's team at the Center for Climate Systems Research, NASA Goddard Institute for Space Studies, publishes monthly compilations of land-based measurements. From its baseline of the average temperature from 1951 to 1980 of 14°C the results are very clear: about 0.4°C of warming from 1900 to 1940, followed by 0.2°C of cooling from 1940 to 1960 and then a somewhat faster warming than earlier amounting to 0.7°C from 1965 to 2011 (Figure 2.4).

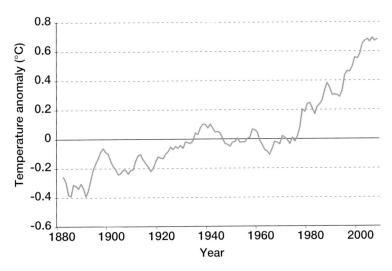

Figure 2.4 The global land-based temperature anomaly (5-year running mean) drawn from data available from the Goddard Institute for Space Studies. The base period for the anomaly (black line) is 1951–80.

These results are for the land, about 29 per cent of the Earth's surface area. Ocean temperatures, more correctly sea-surface temperatures, have been directly measured from ships, using fixed buoys with a thermometer and radio link to satellite and since 1972 by direct satellite measurement of the infra-red radiation from the ocean. Scientists are aware that these three methods do not measure exactly the same thing. A bucket of sea-water hauled aboard a ship may not have come from precisely the same one-metre depth as measured by the fixed buoys, and certainly did not come from the top 0.1 millimetres of the ocean seen by the satellite probe. Recognising and correcting for these differences allows an estimate of sea-surface temperature changes. The next graph (Figure 2.5) is taken from the Intergovernmental Panel on Climate Change (IPCC) Fourth Assessment Report of Working Group 4 (2007, Chapter 3). It shows two of three different analyses of the available sea-surface temperature data, one by the British Meteorological Office (Met Office) Hadley Centre in collaboration with the University of East Anglia Climatic Research Unit (CRU) and one by the Japan Meteorological Agency led by Ishii

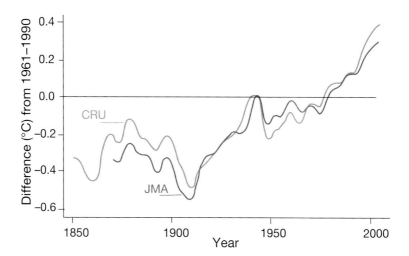

Figure 2.5 Global sea-surface temperature anomalies after the IPCC Fourth Assessment Report of Working Group 4 (2007), Figure 3.4. Two different analyses are shown, one from the British Met Office Hadley Climatic Research Unit in collaboration with the University of East Anglia CRU, the other from the Japan Meteorological Agency (JMA).

(COBE-SST). The third analysis, by the National Climatic Data Center in the United States, pretty much fits between the two illustrated in Figure 2.5. The results from these three disparate groups of scientists are remarkably similar.

Finally in this section, Figure 2.6 uses a world map to show the variation in temperature anomaly across the world for the year 2011. Though warming is evident almost everywhere, the greatest effect is in the northern Arctic.

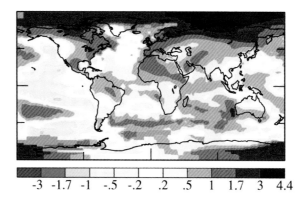

Figure 2.6 World temperature anomaly distribution for 2011. No data are available for areas shaded in grey. Source: Goddard Institute for Space Science Surface Temperature Analysis, NASA.

ANTHROPOGENIC GLOBAL WARMING

You will probably have read and heard the phrase 'anthropogenic global warming' (AGW). It simply means 'global warming caused by the actions of humans'. This term prejudges the issue and I have chosen not to use it. The purpose of this book is to present and consider the science of climate change, including temperature change. Some people do not believe the Earth is warming. Some people consider it probably is, but do not believe human activities are causing it. Others consider that the Earth is warming as a consequence of burning fossil fuels, cement manufacturing and land clearing. When you have read the book you can decide for yourself.

Despite the evidence of extensive compilations of land and ocean records over 130 years or more, sceptics doubt the reliability of the data. Dr Roy Spencer is a former NASA climatologist who for the past 10 years has been based at the University of Alabama. He is prominent on the internet in the debate about climate change, and is particularly sceptical of the land-based estimates of temperature. He argues that many measuring stations are affected by the urban heat-island effect; that is, as population in a town grows, so the surrounds of the meteorological measuring station change, always in the direction of warming. As pasture is replaced by parking lot, as forest is replaced by housing, the net effect is to increase the local temperature. Because of this Spencer considers that significant adjustments need to be made to the published temperature anomalies. Jones and Wigley discuss these matters in a scientific paper published in 2010.[12] They point out that if all the urban temperature measurements are excluded, there is a negligible difference in global temperature estimate and it still shows warming.

But still, doubts remained among a group of physicists from Berkeley University in California, and they decided to make their own independent analysis of the world's climate records. They claimed that they had special expertise in the statistical analysis of data that might be superior to the methods used by the climate scientists. They collected some 39 000 temperature measurements, processed them using their expert understanding of statistics and found exactly the same result as the climate scientists. Including or excluding meteorological stations from cities that might have been biased by the urban heat-island effect made almost no difference to the result.[13] Conclusion: the world is warming.

A different approach to the question of global warming has been to look at the trends in temperature records – records as in 'I just broke the world record'. Whatever the rights and wrongs of the averaging techniques used to establish local temperatures, one might expect the actual measured temperatures to be comparable for that measuring station. If the global temperature is warming, as the years pass it would be less and less likely that record-breaking low temperatures would occur, and more and more likely that record-breaking high temperatures would occur. Such a study by a US research group has found that across the United States, record maxima have become more frequent than record minima over the past 40 years, whereas prior to 1970 the two kinds of records happened with about the same frequency.[14] A similar result can be seen in the Australian Bureau of Meteorology's data, such as its analysis of the percentage of the Australian continent by area that experienced extremely hot years, averaged over successive 40-year periods. The increase since 1960 is very marked (Figure 2.7). There is a similar increase nationally in the average coldest night temperature; data from the Australian Bureau of Meteorology shows that since 1960 there has been a 1.5°C increase.

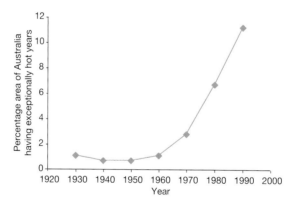

Figure 2.7 Percentage area of Australia experiencing extremely hot years, averaged in 40-year periods; i.e. the point for 1940 represents the years 1920–59. Source: Australian Bureau of Meteorology.

Since 1980, satellite technology has made it possible to determine air temperature at various altitudes. At the lowest level of the atmosphere, climatologists expect temperatures to match surface values. Over the lifetime of the various satellites there have been many analyses and interpretations of the results, some indicating a temperature fall over the period, more indicating a rise. The most recent analyses of all the satellite data indicate a rise of 0.15°C a decade, much as the land and ocean measurements show (see Figure 2.8).

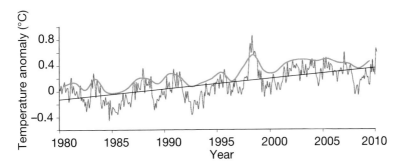

Figure 2.8 Lower atmosphere (troposphere) temperature anomaly derived from microwave sounding units (MSU-TLT) satellite data (zigzag line). Trend of the MSU temperature anomaly = 0.156°C per decade (black line). Global smoothed annual temperature anomaly (HadCRUT3) combined from ocean temperatures compiled by the Hadley Centre of the Met Office and the land-surface temperature records compiled by the Climatic Research Unit of the University of East Anglia (smooth line). MSU data are produced by remote sensing systems and sponsored by the US National Ocean and Atmospheric Administration (NOAA) Climate and Global Change Program.

We saw in Chapter 1 the view of those who deny that global warming is even happening. Their main argument is that the world has cooled since 1998. Every month the US National Climatic Data Center reports the world's average temperature, considering land-based measurements, sea-surface temperatures and the global average. The warmest years on record were 2005 and 2010; both years were 0.64°C above the 20th-century average. Slight cooling during the strong *La Niña* of 2011 dropped the average temperature to 0.51°C above the 20th-century average.

SUMMARY

Topic	Observation	Key statistic	Conclusion
Spring timing	Earlier	Up to one month	Warming
Winters	Milder	Australia's average coldest temperature has risen 1.5°C since 1960	Warming
Land temperature	Rising	~1°C in 100 years	Warming
Sea-surface temperature	Rising	0.8°C in 100 years	Warming

Topic	Observation	Key statistic	Conclusion
Atmospheric temperature	Rising	0.5°C in 30 years	Warming
Extreme temperature events – record maxima	More frequent	March to June 2010 broke all previous records	Still warming

Do we now have an answer to the question: Is the climate changing? Not entirely. Although we have seen that over the past 130 years there has been a steady increase in the global temperature, amounting to 1°C, temperature is but one aspect of climate. We have also seen that the temperature rise has fluctuated. For example. there was a hiatus in the warming between 1940 and 1965 and a sharp rise in 1998. Sceptics point to this and other variations to argue that it cannot be carbon dioxide that has its finger on the thermostat because the atmospheric carbon dioxide content does not vary in that way. So what does cause the apparently random ups and downs in the temperature graph? They would be more properly called changes in the global weather rather than in the climate, but both are inextricably linked, and in the next chapter we will look at what makes weather and what makes climate.

FURTHER READING

Hansen J (2009) *Storms of My Grandchildren*, Bloomsbury.

3

WEATHER IS NOT CLIMATE

Some are weatherwise, some are otherwise.

Benjamin Franklin

'Weather' is what happened today or yesterday, or this year, or since the baby was born. 'Climate' is what you understand about a place when you have lived in it for 30 years or longer. If the place was Adelaide, you may remember hot summer days occasionally reaching 40°C, frosty winter mornings and cold school rooms, but mostly an equable, so-called Mediterranean climate. If you lived in Scotland perhaps you recall freezing winters with a metre of snow, chilly springs with gusts of rain and so on. The weather varies very widely wherever you are, but the climate, well, it really stays the same over half a lifetime – or at least it used to, because climate is the average of all the weather's variations.

During most of the first decade of the 21st century, southern and western Australia experienced severe drought. Explosive bushfires in the summer of 2008–9, the drying of the Murray and the Darling rivers, unheralded water restrictions; to those younger than age 60 these seemed to be a remarkable change from the normal climate. But there were enormous fires in Victoria in 1939, there were devastating droughts at the end of the 19th century and before the construction of dams and weirs the Murray would shrink during drought to a series of pools (see Figure 3.1). Droughts (and flooding rains) are just part of the Australian climate. Particularly long droughts, those lasting for up to 10 years, occurred in the periods 1898–1905, 1911–16, 1939–45, 1958–68 and 1991–96.

To go outside on a hot summer's day with no rain in sight and having had none for the previous three months, and then to attribute that weather to climate change is to misunderstand the issue. It is certainly possible that the climate could change to make drought in southern Australia more intense, just as it is possible the rainfall across northern Australia could increase, but local experience provides no argument for it. Equally, a hot summer or a cold winter does not provide an argument for or against global warming. If you live in Beaufort in Victoria you might well believe that the climate has not changed in the past 50 years, and going by the rainfall, you would be right for that locality. If you live in Denmark in south-western Western Australia, you might worry a great deal about the declining rainfall and attribute this to climate change. Both attitudes could be wrong because they are based on local and personal experience.

The science of climate change must depend on accurate and extensive knowledge of weather in all its variations, of local averages and of worldwide weather phenomena and their causes. For by its very definition, we cannot even recognise climate *change* if we have no knowledge of climate history on short (100-year), medium (1000-year) and long

Figure 3.1 Murray River at Mildura, probably during the drought of 1901. Source: State Library of Victoria, Wilf Henty, photographer.

(100 000-year) time scales. In this chapter I will describe some of the science that provide the basis for being able to assess whether change is happening.

SUNSHINE

Since climate is the average of weather taken over the years, whatever causes the weather also causes the climate. So next we will look at the weather engine, the Sun, but first I want to say something about light and heat.

Sunshine is made up of heat, light and ultraviolet (UV) rays, with most of its energy coming in as light, the colours of the rainbow. Like radio waves, which can be short waves or long waves, the different parts of sunshine can be described as having different wavelengths. UV has the shortest wavelength and is the most energetic. Heat – also called infra-red – has longer wavelengths than UV. All of them, heat, light and UV rays, warm the planet. You cannot specifically 'feel' UV, but it can burn your skin nonetheless.

Some of the incoming energy from sunshine is absorbed in the air; ozone located high in the atmosphere takes out a lot of the UV, and water vapour absorbs some of the Sun's heat energy. Figure 3.2 shows the full

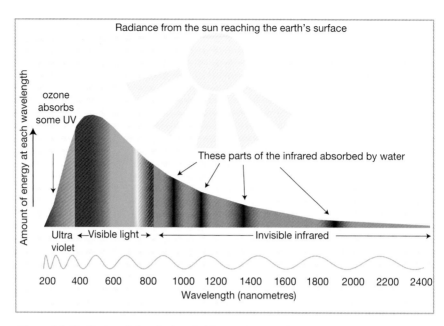

Figure 3.2 The Sun's radiation. Each part of the spectrum, not just the visible part as we see it in a rainbow, has its own wavelength, measured in nanometres (millionths of a millimetre). Various gases in the atmosphere, particularly ozone and water vapour, absorb different parts of the Sun's radiance, as shown by the dark bands.

range of the Sun's radiation spectrum, with dark bands indicating where ozone or water vapour absorb some of it.

The Earth is a spinning planet illuminated by the Sun. The Sun provides the energy to drive the weather, and from satellite measurements we know there is over a kilowatt (kW) of energy falling on each square metre when the Sun is directly overhead – enough to power a small radiator. By the time it has passed through the atmosphere, that energy is reduced to a bit less than a kW during the Australian summer, and half a kW or less at noontime in winter.

MILANKOVIĆ CYCLES

The Earth's aspect to the Sun changes over time, and as it does the impacts of summer and winter change. If the Sun's radiation pattern changes; for example, northern hemisphere summers grow colder because the Earth is a little further away, less polar ice melts (see next page). The pattern of the Sun's energy reaching the Earth depends on slow and small changes in the Earth's orbit. These changes in the Sun's energy amount locally and seasonally to as much as 100 watts per square metre and they vary at a rate measured in thousands of years. The orbital changes were first mathematically combined by Milutin Milanković in 1920 and used to explain the ice ages that the Earth periodically experienced from about 400000 years ago until 10000 years ago, as shown in Figure 3.4 (p. 31). During those ice ages, the Earth cooled and warmed four times, with glaciers advancing during the cooler periods. The last glaciation ended about 10000 years ago, and during that period global temperatures were about 6°C lower than today. There was then, geologically speaking, a very rapid temperature rise at a rate of about 0.5°C per 1000 years to reach today's temperatures.

MILANKOVIĆ CYCLES

The Milanković cycles refer to periodic variations in the amount of sunshine reaching particular regions of the Earth. Because most of the Earth's land is in the northern hemisphere, that is the region usually considered. The factors that cause the variations are astronomical, relating to the Earth and Sun. The Earth rotates once a day about its axis, and rotates once a year about the Sun. The Earth's rotation axis, North Pole to South Pole, is currently tilted 23 degrees away from vertical with respect to its orbit around the Sun (Figure 3.3a). This tilt, in combination with the Earth's orbit around the Sun, is what makes the seasons. In Figure 3.3a, the Earth is shown at summer in the northern hemisphere and winter in the southern hemisphere. Six months later the Earth will be out to the left of the Sun and the seasons will be opposite. When the Sun is more overhead at midday it is summer, when the Sun is lower at midday it is winter.

The Earth's orbit is not a circle, but an ellipse, or an oval, and the distance the Earth is from the sun changes during the year. Figure 3.3b, much exaggerated with respect to the change in the Earth–Sun distances, shows the position of the Sun and Earth when the Earth is closest to the Sun during summer in the northern hemisphere, making these summers warmer than if the opposite were true (Figure 3.3d). Because the summers are warm, winter snow melts and the ice does not accumulate very far south of the North Pole, leading to an interglacial. The shape of the Earth's orbit changes on an approximately 100000-year cycle, so that 50000 years after Figure 3.3b, something like Figure 3.3d happens. The northern summers are cooler, less ice melts, the winters are cooler, more ice forms and this leads to glaciation. Exactly the opposite applies to the southern hemisphere, but there is only one land mass anywhere near the South Pole, so there is nowhere other than Antarctica for ice to accumulate. As far as Antarctica is concerned, it is still an ice age!

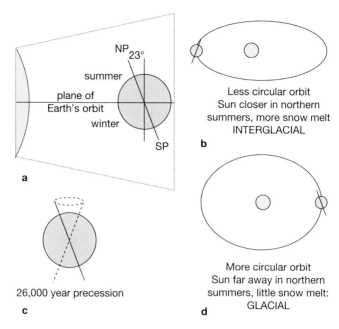

Figure 3.3 Depiction of some of the Earth's orbital parameters that affect climate.

Coupled with the change in orbital shape is a small change in the tilt of the Earth's axis, from 22.1 to 24.5 degrees. When the tilt is less, summers are cooler and winters warmer. Adding this variation complicates the 100 000-year cycle, which is further complicated by the 26 000-year precession cycle of the Earth's axis tilt (c). Together, these astronomical factors make the Milanković cycles (Figure 3.4), where the wavy line shows how the amount of energy from the Sun striking 65° North latitude has changed over the past 400 000 years.

The periodic variation in solar energy on time scales of thousands of years is as well understood as the reasons for seasonal weather variations. The last peak in the Sun's heat was 6000 years ago and we are now on a very slow cooling phase; so slow that the next ice age is not expected for about 50 000 years.

SUNSPOTS AND THE SUN'S INTENSITY

A much quicker, though very much smaller source of variation in the Sun's energy is the variation in the number of sunspots, which are strong magnetic storms on the Sun. Though these dark spots send less energy our way than would arrive in their absence, they are associated with bright

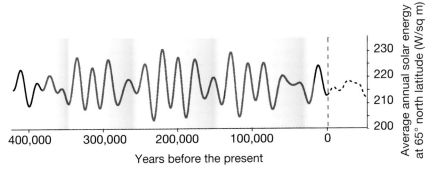

Figure 3.4 Milanković solar cycles for the region north of 65° N latitude over the past 400 000 years, with four ice ages shown in shaded blue – darker is colder. Note that whenever there is a prolonged period without higher peaks of solar energy, there is an ice age. The future is predicted by the dashed black line; this is quite similar to the conditions that started the last ice-age, about 70 000 years ago. Source: www.climatedata.info.

patches called 'faculae', which send more energy, and this outweighs the sunspots effect. Consequently there is an increase in solar energy when there are many sunspots. Sunspot numbers increase and decrease in an approximately 11-year cycle. The last sunspot maximum was in 2000–1; 2009 was a minimum and the return to the expected 2012–13 maximum began early in 2010.

SUNSPOT CYCLE

The change in the Sun's energy through each sunspot cycle of 11 years depends on the number of spots. In some cycles no spots at all are seen, and when low numbers continue for several 11-year cycles there is a detectable cooling of the Earth (Figure 3.5). A pronounced absence of sunspots from 1645 to 1715 is called the Maunder Minimum and a less marked reduction from 1800 to 1840 is called the Dalton Minimum. Old records of sunspot numbers are fascinating for they give names of observers through history such as Galileo, Cassini, Boyle, Huygens, Hooke and Halley.

As the sunspot numbers rise, so does the Sun's energy output, as shown in the paired graphs of Figure 3.7, where sunspot numbers are shown by the dark line and variation in solar energy by the pale line. While the graph clearly shows there is a relationship, it exaggerates the energy effect, because the range of variation in solar energy is only from 1365 to 1367 watts per square metre; that is, the variation is about one-tenth of one per

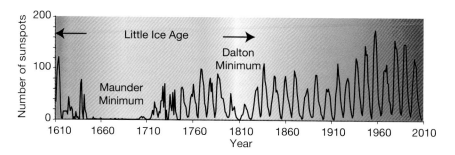

Figure 3.5 Sunspot cycles over the past 400 years.[1] Cooler periods are shaded in blue, warmer periods in red.

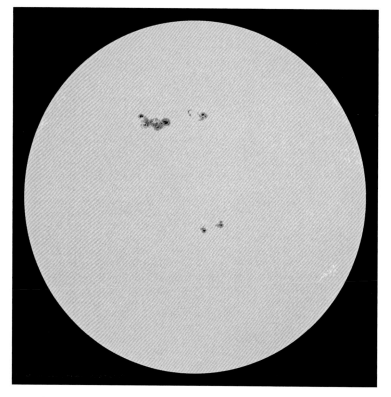

Figure 3.6 Sunspots photographed in March 2012. Source: Solar and Heliospheric Observatory, European Space Agency and NASA.

cent. Small though this amount is, it does affect the Earth. Small organisms called diatoms living in lakes need light and warmth to grow, and they appear to increase in number in step with the increase in sunspot numbers. This conclusion is reached by measuring the thickness of mud deposited in the lakes– mud that includes the bodies or shells of these

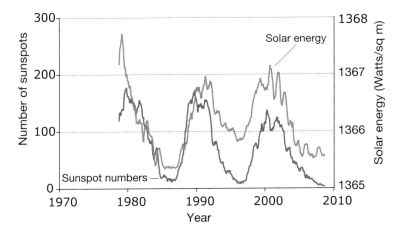

Figure 3.7 Variation in the Sun's energy as the sunspot number varies. Solar energy variation measured by satellite is the upper line, while sunspot numbers are shown by the lower, darker line. Source: NOAA.

organisms.[2] When the Sun's output is higher, the diatoms, which depend on sunlight for their growth, thrive more abundantly. They only have a short life span and since in such a year more grow, by the end of that year more dead bodies have accumulated on the lake floor than in a year when the sun's output is lower. Hence, years with many sunspots produce thicker layers of diatom sediment.

Figure 3.8 A variety of freshwater diatoms seen through a microscope. The larger diatoms are about one-fiftieth of a millimetre across. Source: Micrographia.com.

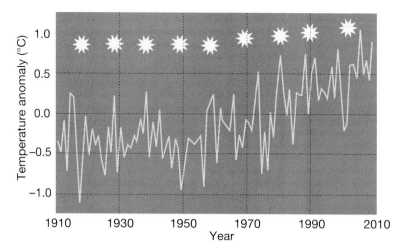

Figure 3.9 Australian mean temperature anomalies (difference from the 1961–90 average). Sunspot maxima are indicated by the Sun cartoons. Source: Australian Bureau of Meteorology.

Some of the irregularities in the Australian temperature graph shown again here in Figure 3.9 seem to coincide with sunspot maxima and minima. Each 11-year sunspot maximum is indicated by a Sun. Early in the century there is a good match, then not in 1948, then matching fairly well up until 2001, a year of a strong, cooling *La Niña*. Already some of those temperature oscillations might be at least partially explained.

There is a longer-term change in the Sun's strength that has been analysed by various groups of astronomers, using historical records of sunspot counts and proxies going back 1000 years. These show minor fluctuations in the Sun's intensity, cycling over periods of 200 or 300 years, with an overall increase since 1850 of less than one-tenth of one per cent. According to Professor Lockwood of the University of Reading in the United Kingdom, since about 1980 there may have been a very small overall decrease in the Sun's irradiance.[3] However, there are some who argue there has been a small increase in the Sun's irradiance and that this change has contributed up to 70 per cent of recent global warming.[4] Others attribute no more than 30 per cent of the past 40 years of global temperature increase to the Sun.[5] Most recently a review article by Judith Lean of the US Naval Research Laboratory concluded that the 11-year sunspot cycle causes a temperature fluctuation of about 0.1°C, and that since 1850 a gradual overall increase in the Sun's irradiance has added perhaps another 0.1°C to the global temperature.[6] This topic enjoys considerable debate and while it is certainly not yet settled, most of those investigating the Sun's contribution to global warming since 1850 seem to lean towards a figure of about 10 per cent. The observed temperature change since 1850 of about 1°C has not been caused by the Sun.

EXTRA-TERRESTRIAL EFFECTS

Impacts on the Earth's atmosphere, and as a result on climate, can come from even further away than the Sun. We are continually exposed to cosmic rays: high-energy radiation, higher energy than X-rays, higher even than comes from atomic bombs. Luckily there is not much of this type of radiation, and it has been coming here forever, so unless you are an astronaut there is no reason to worry. Most of these cosmic rays come from the Sun, while some come from far beyond. A few astronomers and earth scientists believe that changes in the quantity of cosmic rays cause changes in climate. One such is Professor Svensmark from Denmark, who has analysed the Earth's climate history over the past 500 million years and the influx of cosmic rays. He sees a connection between periods of high Earth temperature and low cosmic rays, and vice-versa.[7] One claimed effect of cosmic rays is to enhance the condensation of water vapour into water droplets – clouds. When cosmic rays are more intense there will be more clouds, and since clouds reflect some of the Sun's heat, more clouds means a cooler Earth. Tests of this theory by comparing cloud cover and cosmic radiation in recent times have failed to support it.[8] One study in Finland over the years 1996–2008 examined sudden bursts of cosmic rays and found they played at most a minor role in aerosol production and subsequent cloud cover.[9] Another study concluded that the hypothesised effect is too small to play a significant role in current climate change.[10]

THE ATMOSPHERE

If the brightness of the Sun were the only factor controlling the planet's temperature, we would, like the Moon, have below-freezing surface temperatures averaging minus 20°C. Today, the global average temperature on Earth is about 14°C. The difference is because the Moon has no atmosphere. It is our atmosphere that keeps the Earth at a habitable temperature. Apart from its obvious usefulness in providing some air to breathe, the atmosphere does a critical job as an absorber of the Sun's radiation and as a 'blanket' around the Earth.

If the Sun's energy does not penetrate the Earth, it cannot warm the surface. We saw earlier how the atmosphere blocks some of sunshine, with ozone filtering much of the lethal UV, and water vapour cutting some of the infra-red. Visible light fits in between those two invisible parts of the Sun's radiation spectrum, and almost all of the light gets through. It is reduced somewhat by what is known as Rayleigh scattering; the blue-sky phenomenon, which is the scattering of sunlight by the molecules of the air. Some of the scattered light goes back into space, thereby reducing the Sun's input. Blue light scatters more than red, so the light that reaches your eye after being scattered is richer in the blue end of the spectrum, whereas the direct light from the Sun, having not been scattered much, still contains the yellower part of the spectrum. At sunset you see more of the colour range as the direct path to the setting Sun passes through so much more air. Rayleigh scattering does reduce the incident sunlight, but it does so uniformly and is probably not a factor in climate variability.

AEROSOLS

Bigger particles than the molecules of oxygen and nitrogen lurk in the upper atmosphere, some a consequence of fossil-fuel burning. These are known as aerosols. Sulfate is one; either in the form of tiny droplets of sulfuric acid or as particles of ammonium or other sulfates. Sulfate comes both from the burning of sulfur-containing fuels and from volcanoes, and has a net cooling effect because it partially reflects sunlight. Larger particles still are volcanic dust and ash as well as soot from burning. The bigger the particle the shorter the time it stays in the air before falling or being rained out. Fine volcanic dust certainly hangs about for several years, as testified by red sunsets for a year after the explosion of Krakatoa in 1883.

Soot, also known as black carbon, gets into the air from almost any-thing that burns, especially forest fires and wood-burning cooking fires, but coal, oil, diesel and petrol all contribute. The overall impact of soot is one of the lesser-known parts of the climate story. Being black, soot absorbs radiation so its effect in the atmosphere is to warm it; soot does not reflect much sunlight in the way that sulfate aerosol does. When soot falls out of the air it can darken the Earth's surface, especially if it falls on snow. Some studies suggest that the Himalayan glaciers are melting at an increased rate because there is a veneer of soot over the ice. Shrestha and colleagues recently published a review about this material; they consider that black carbon could rank second to carbon dioxide as a contributor to global warming.[11]

An analysis of solar radiation since 1950 published by Ohmura[12] in 2006 found clear evidence of a steady dimming of the Sun from 1950 up until about 1980, attributed to the rising presence of reflecting aerosols in sulfurous smoke as industry recovered after the Second World War. Then, following the 1979 Geneva Convention, at which there was as international agreement to cut atmospheric pollutants, there was a decline in aerosols and an increase in the amount of sunshine reaching the surface. In 2009, Philipona and colleagues[13] claimed that 60 per cent of the 1°C rise in temperature over mainland Europe since 1980 was the direct result of the reduction in aerosol content of the European atmosphere.

The global temperature anomaly as combined from ocean temperatures compiled by the Hadley Centre of the UK Met Office and the land surface-temperature records compiled by the Climatic Research Unit of the University of East Anglia is shown in Figure 3.10. The sunspot periodicity fits several of the temperature peaks and the aerosols pattern appears to contribute to some of the troughs. Earlier aerosol data are only available for Europe; they show a dimming of the received sunshine from 1940 until 1975 followed by brightening up to 2005. The short, sharp temperature drops in 1963, 1982 and in 1991–92 are attributed to world-wide clouds of volcanic dust after the eruptions of Mt Agung, El Chichon and Mt Pinatubo respectively. Add in the eruption of Krakatoa in 1883 and some, though by no means all, of the variation in the temperature graph is explained.

The particular impact of aerosols is relatively short term, affecting temperatures over periods of one to five years. There are other phenomena

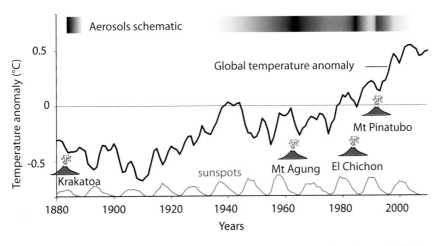

Figure 3.10 Global temperature anomaly since 1880 with major aerosol and sunspot effects noted. Sources: Hadley Climatic Research Unit (HadCRUT3, temperature data) and Hoyt & Schatten, 1998 (sunspot curve).[1]

that are there all the time. As well as dust and aerosols periodically blocking some sunlight, the atmospheric gases themselves absorb some of the sunshine before it reaches the surface of the Earth. Specifically in the infra-red, or heat, part of the spectrum, there is energy absorption by water vapour and carbon dioxide. Absorption means giving energy from the radiation to a molecule of the gas, be it ozone or water vapour or carbon dioxide. Through absorption the Sun's radiation energy reaching the surface is decreased but the atmosphere gets warmer. So while some aerosols act to cool the Earth by reflecting sunlight, others such as soot cause warming; and the absorption of heat by the air itself serves only to heat the atmosphere.

REFLECTION

So far we have only considered the Sun's radiation that reaches the Earth's surface. Radiation does not all stay at the surface. About 30 per cent of it is reflected straight back out into space (see Figure 3.11). Clouds and snow are the best reflectors, sending back into space more than three-quarters of the sunshine that hits them. The world's desert regions reflect about one-third of the sunshine that falls on them, grasslands and forests rather less and the ocean absorbs almost all the heat it receives from the Sun. It is probably lucky that there is so much ocean, otherwise the Earth might be a great deal cooler.

That all adds up to tell us that the first part of the Earth's energy budget, the incoming heat from the Sun, can change in four main ways:

1) by the sun itself changing

2) through the small changes in the Earth's orbit, the Milanković cycles

3) by aerosols dimming the sunshine

4) by the Earth reflecting more or less of the sunlight back out.

The second part in the Earth's energy budget is about how some of its own heat radiation gets blocked, or absorbed, by the atmosphere and cannot get away.

ALBEDO

Derived from the Latin word *albedus,* meaning 'whiteness', the albedo of a planet is the fraction of the Sun's energy it reflects back into space. Venus, the morning and evening star, is bright because it has a very high albedo. Sulfuric acid clouds cover the whole planet and they reflect 65 per cent of the Sun's light, thus its albedo is 0.65. Earth's albedo is 0.30, and the Moon's 0.12. Different surfaces on the Earth have their own albedo, shown in Figure 3.11.

Figure 3.11 Percentage reflection of sunlight by various terrains on Earth. All together the Earth reflects 30 per cent of the Sun's radiation back into space.

THE ATMOSPHERE AND THE OCEAN

The vital ingredients that actually make the climate are the atmosphere and ocean. Where there is no atmosphere there is no climate and no weather. Most importantly, as mentioned earlier, the atmosphere is the Earth's 'blanket'. The Sun warms the Earth, but its warmth would all be radiated back to space if the atmosphere did not keep at least some of that warmth in. This is a topic that is described in considerable detail in Chapter 4.

The atmosphere is also important in the way it moves heat around. One look at a weather map as a cyclone approaches shows where most of the weather comes from; it comes from the oceans. Oceans cover 70 per cent of the planet and they store vastly more heat than the atmosphere does. The Sun's energy evaporates water from the oceans and the warm, humid air rises. As it does, it cools and the water vapour changes back into water droplets (clouds). This is the Earth's heat exchanger; energy from the Sun is used to turn water into vapour, and this energy is returned into the atmosphere when the vapour condenses back to water. The added energy makes the air move faster and the Earth's rotation generates the spiral patterns that characterise not only cyclones but also all of the atmosphere's wind directions.

Ocean currents provide the major exchange of heat between the ocean and atmosphere. Foremost among them is a global current known as the Great Ocean Conveyor Belt. Cold, salty, dense polar water sinks to the ocean floor and slowly makes its way towards the equator. It is replaced by warmer tropical water, which being less dense, flows at the surface (see Figure 3.12). The presence of the continents pushes these currents in various directions so their final paths are quite complex, but the end result is to move heat from the tropics to the Arctic and the Antarctic regions. Probably the best known of these is the Gulf Stream, which moves warm water from the Gulf of Mexico northward across the Atlantic towards western Europe. The Antarctic Circumpolar Current is the largest and strongest part of the Great Ocean Conveyor Belt.

The Humboldt Current is one of the most important players, for it contributes to the *El Niño* and *La Niña* phenomena. This is a complex series of events across the tropical Pacific, ruled by the easterly trade winds. When they are mild and constant (situation normal, Figure 3.13a, p. 43), warm surface water moves westward and cool, nutrient-rich water rises from the Humboldt Current off Peru on the South American continent. The fishing there is good. When the trade winds ease or are inconsistent,

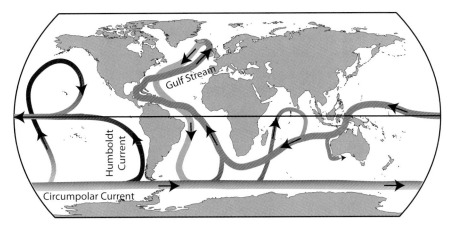

Figure 3.12 Main features of the Great Ocean Conveyor Belt. Deep, colder currents are shown in blue, surface, warmer currents are in orange and intermediate temperatures are indicated in magenta. Note particularly the northward Humboldt Current, which carries cool Antarctic water up the South American coast before it turns west and is warmed. Note also the warm surface current crossing the north Atlantic Ocean from the Gulf of Mexico to warm waters around Britain. Source: *World Ocean Review 2010*, Maribus, Hamburg.

the warm surface water stays in the east, blocking the rise of the cooler water (see Figure 3.14). The fishing is poor, and as this normally happens around Christmas, Peruvian fishermen call the warm current *El Niño*, which means 'the Christ child' (Figure 3.13b). Should the trade winds grow stronger than normal, the warm water is pushed far to the west, reaching Australia, and because the situation is the opposite of *El Niño*, it is called *La Niña*, the girl child (Figure 3.13c). Each phase in the transition from normal to *El Niño* to normal to *La Niña* may last a year or two, and the repeated swings have become known as the Southern Oscillation. Combining these names leads to the meteorological term *El Niño* Southern Oscillation (ENSO).

As you will know from watching weather maps, high-pressure systems bring dry, warmer weather, while lows bring the rain. As the warm Pacific water heats the air above it, the air rises, carrying evaporated water with it, and because the air is rising it produces a low-pressure region. The winds move this moist air and low-pressure region further west, and as the air rises it cools, the water vapour condenses and it rains. During *La Niña* the moist air drops rain over the western Pacific – over eastern Australia and Indonesia – leading to monsoonal rains and possibly floods. During *El Niño* the rain falls out before reaching the land masses, and eastern Australia experiences drought.

The atmospheric circulation pattern resulting from the Pacific sea-surface temperature variation is called the Walker Circulation, named after a British meteorologist who first recognised that air-pressure differences across the Pacific were related to atmospheric convections cells that affected the tropical monsoon. During *El Niño* higher air pressure over Australia brings warm, dry summers, while low air pressure during *La Niña* brings the humid, wet summers. In south-western United States the opposite holds. The atmospheric pressure difference between Tahiti and Darwin is called the Southern Oscillation Index, or SOI, and is used by meteorologists to indicate the strength of *El Niño* (negative) or *La Niña* (positive).

Just as *El Niño* and *La Niña* affect Australia's weather, so do ocean currents in the north Atlantic affect the weather in Europe, North America and the Arctic. Among these is the Atlantic Multidecadal Oscillation, a cyclical change in sea-surface temperature that varies by about half a degree every 60 to 80 years (Figure 3.14, p. 45).[14] Michael Schlesinger and Navin Ramankutty's 2011 analysis of this climate influence concluded that it was not a worldwide phenomenon, though its influence on European and North American temperatures made it a significant factor in estimates of global temperature variation.[15] If such a 60-year cycle is real, natural and affects the climate, it should be evident in temperature reconstructions for before 1900. One analysis of climate proxies for the north Atlantic over the past 8000 years found a quasi 50–75-year cycle,[16] as did another analysis covering the past 500 years.[17] A scientific paper by Vinczi & Jánosi entitled 'Is the Atlantic Multidecadal Oscillation (AMO) a statistical phantom?' concluded that it was not regular and should be called the Atlantic Multidecadal Variation.[18] A summary of this topic by Richard Kerr in 2012, quoting Michael Mann, concluded that '… the evidence for this sort of 50–70 year oscillation is accumulating'.[19]

The Atlantic picture is complicated by a pressure oscillation similar to the SOI, called the North Atlantic Oscillation, and proxies show it has been an important climate factor. Some find a 60-year cycle,[20] while others[21] find variation but no regularity. To the extent that such pressure and temperature changes are linked, they help to explain climate variability over the region of the northern Atlantic.

Research by Shoshiro Minobe from Hokkaido University also finds a 50–60-year cycle in sea-surface temperatures, but in the north Pacific and Indian oceans.[22] Temperature minima appear around 1910 and 1960 in three areas, a little different from those of Figure 3.15, but in the fourth these years are temperature maxima. Minobe concludes: 'the 50–70 year

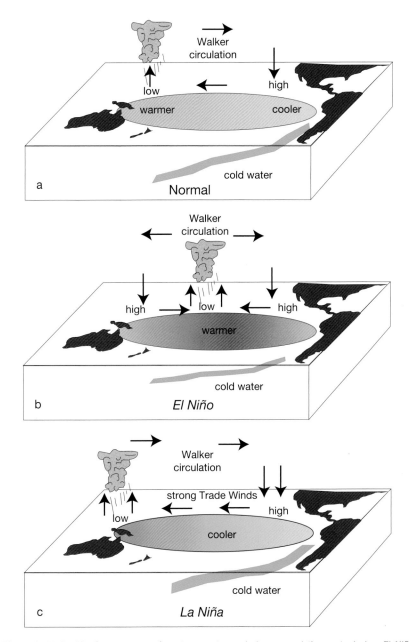

Figure 3.13 Pacific Ocean sea-surface temperatures during normal times a), during *El Niño* b) and during *La Niña* c). Blues are cooler, orange and red warmer. Different Walker circulation patterns also distinguish normal times from *El Niño* or *La Niña*.

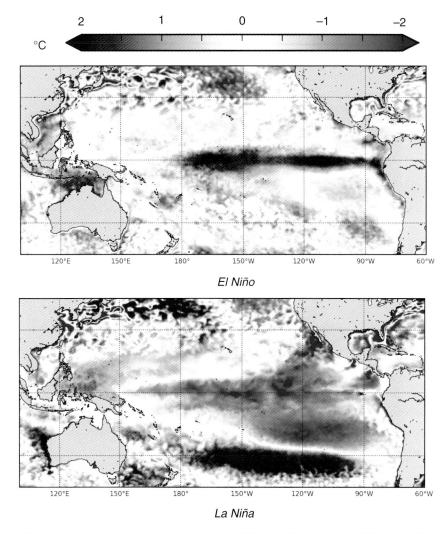

Figure 3.14 Sea-surface temperatures during an *El Niño* event (above) and *La Niña* (below). The temperature difference between the two phases is very obvious. Source: Australian Bureau of Meteorology, products of the Pacific Islands Climate Prediction Project.

variability is likely to be essentially an internal oscillation in the coupled atmosphere–ocean system'. A set of sea-surface temperatures from the tropical South Pacific Ocean spanning the past 270 years was compiled by Braddock Linsley and colleagues, using corals as the proxy.[23] They found general cooling over the period, with fluctuations of the order of 0.75 degrees at a frequency of around 10 years. The fluctuations approximately match those in ENSO and the Pacific Decadal Oscillation but not a 60-year cycle.

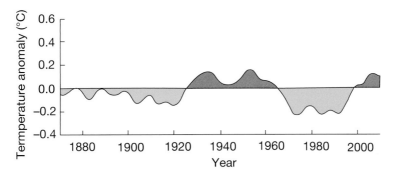

Figure 3.15 Sea-surface temperature changes of the Atlantic Multidecadal Oscillation, after Trenberth & Shea (2006).

Other oceanic weather patterns are important to other parts of the world. For eastern Africa and Indonesia the so-called Indian Ocean Dipole, which refers to sea-surface temperature differences across the Indian Ocean, particularly affects rainfall and is thought to be a contributor to drought in southern Australia.[24]

From these studies it seems that there is no doubt variation in climate, at a scale of a few years (*El Niño– La Niña*) or a few decades (Atlantic Multidecadal Oscillation), is caused by ocean–atmosphere interactions. It is also evident that much as we would like to, we do not yet understand these forces well enough to use them to predict climate more than a short time (months) ahead.

THE ANTARCTIC

Another contributor to the global climate is the vast Antarctic continent, covered as it is by up to a depth of 5 kilometres of ice. An extensive article by Mayewski and 17 co-authors from seven different countries details the way in which the Southern Ocean and Antarctica affect the whole world's climate.[25] In this work, among many other things, the authors examine Antarctic and Southern Ocean climate and its relation to the global climate system. Of this they write:

> A comparison of reconstructions of Northern and Southern Hemisphere temperature … covering the last 2000 years … demonstrates that major changes in temperature and circulation intensity are associated, such that cooler temperatures coincide with more intense atmospheric circulation and warmer

temperatures with milder circulation. Further, until the warming of the last few decades, major changes in temperature were preceded by or coincident with changes in atmospheric circulation. Modern warming is not preceded by or coincident with change in atmospheric circulation, suggesting that recent warming is not operating in accordance with the natural variability of the last 2000 years and that therefore, modern warming is a consequence of non-natural (anthropogenic) forcing.[25]

THE ATMOSPHERE AND THE LAND

Could some of the 'non-natural forcing' have come from the way we use the land? It is thought that farming as we know it – the clearing of original vegetation and the planting or encouragement of 'useful' crops – has been going on since the end of the last ice age, when half the Earth was covered by forest.[26] Using such disparate clues as pollen counts in lake sediments, evidence from archaeological digs and historical records, it is estimated that since the Industrial Revolution there has been three times as much forest cleared as in the 10000 years prior.[27] Now, we are down to 30 per cent forest cover.

Farming changes the landscape, and any landscape change has, literally, flow-on effects. Clouds are not only made from water evaporated from the oceans. The land continuously gives water to the atmosphere, predominantly by plants breathing, or transpiring as it is called, water vapour from their leaves. Tropical rainforests are the biggest land-based source of water vapour, over which there is a degree of cycling of the water, as the transpired water vapour falls back as rain. When forests are cleared for cropping, the vapour flow to the atmosphere is greatly reduced. In Australia this has led to widespread salinisation of the soil, for when deep-rooted plants are removed, the water table rises, bringing denser, salty water near the surface and in low areas, right to the surface. It is thought that land clearing in the sub-Saharan region will have an impact on the monsoon in tropical Asia as the African contribution to the atmosphere of the Indian Ocean basin declines.[28]

Just how much the Australian landscape and climate have changed through the impact of people is a much-discussed issue. One suggestion is that systematic burning over the past 60000 years modified the landscape from tree and shrub to desert scrub, and this in turn affected the south-ward penetration of the tropical monsoon, changing the rainfall pattern.[29] However, a later study found '… the role of vegetation change in explaining the monsoon system is negligible'.[30] A group of Queensland-based scientists investigated the effect on climate of land clearing since European settlement, using computer modelling.[31] They found that in south-western Western Australia and in eastern Australia there has been statistically significant warming, particularly during drought, as a result of land clearing.

The type of vegetation cover also affects climate in the way it reflects the Sun's heat. North of the Arctic Circle, during the winter, cleared land is snow-covered and reflects the Sun's heat much more than forest does. Here, clearing has an overall cooling effect. By contrast, in Amazonia, clearing forest to pasture creates a warmer, drier climate.[32]

TEMPERATURE FLUCTUATIONS

To the variation in global temperature caused by sunspots and aerosols we can now add the stronger ocean influences, ENSO and the Atlantic Multidecadal Oscillation. Combining these factors on one diagram reveals much of the reason the temperature graph has so many fluctuations (Figure 3.16).

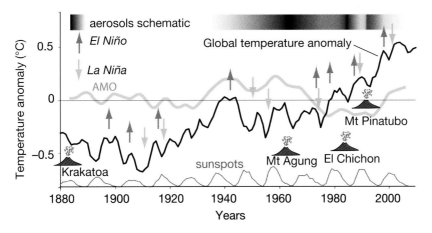

Figure 3.16 Global temperature variation (black line) with aerosol impact, the Atlantic Multidecadal Oscillation (green line), warming *El Niño* (upward arrow), cooling *La Niña* (downward arrow), major volcanoes and the sunspot cycle (lower, pale curve). Sources: Hadley Climatic Research Unit (HadCRUT3, temperature data) and Hoyt & Schatten, 1998 (sunsport curve).[1]

Most of the irregularities in the temperature curve coincide with one or other of these climate influences. They explain why the global temperature goes up and down by a few tenths of a degree over periods of several years. Many of these irregularities can be seen as two or three-year long mild changes in the local weather, such as an usually wet winter or a few years of drought. What none of these influences on the climate can explain is the overall upward climb of the temperature since 1750.

When taken together over decades or centuries, the heat from the Sun, the ocean currents, atmospheric dust and the winds, all combine to make the climate. They make a desert across the Sahara and central Australia, jungles in Borneo and the Amazon, prairies in Canada, steppes in Russia. They make Liverpool in western England warmer today than St John's, Newfoundland, though Liverpool is 6 degrees nearer the North Pole. Climate is where the winds blow, where the ocean currents run and where the rains fall.

SUMMARY

Topic	Observation	Key statistic	Conclusion
Sunshine and the Milanković cycles	Explains the ice ages up to 400 000 years ago.	The last peak in temperatures was 6000 years ago and the next ice age should be in around 50 000 years.	The world *should* be in a very slight natural cooling cycle.
Sunspots	Increase in solar energy due to spots on the Sun in an 11-year cycle.	May have contributed to a 0.1°C increase in temperatures since 1850.	The science is not settled. The majority of scientific evidence indicates that sunspots have not significantly influenced long-term temperature change.
Cosmic sources	Might affect clouds.	Inconsistent	Probably no impact on climate change.

Topic	Observation	Key statistic	Conclusion
The atmosphere – aerosols	Like shade-cloth, dims the Sun.	Industrial aerosols from 1950 to 1980 held European temperature rise back by half a degree.	Increasing aerosols causes cooling.
The atmosphere – reflection	Reflects the Sun's heat.	Only 70% of solar radiation warms the Earth.	Changing cloud cover influences surface temperatures.
The atmosphere and the ocean	Transfers heat from ocean to atmosphere and vice versa, and throughout the oceans.	Variable on a scale of several years to decades	Ocean currents strongly influence climate and weather.
El Niño/ La Niña	Varies irregularly on a scale of 3 to 5 years.	Minor contribution to global temperature variability.	Most influential ocean phenomenon

Temperature change drives climate change. If the climate changes then the weather changes, and when the weather changes our lives change. Global warming is attributed to the greenhouse effect of carbon dioxide added to the atmosphere since 1750. How can adding a little carbon dioxide to the atmosphere change the whole world's climate? Read on!

FURTHER READING

El Niño Southern Oscillation (ENSO): www.bom.gov.au/watl/about-weather-and-climate/australian-climate-influences.shtml?bookmark=enso

Martin JE (2012) *Introduction to Weather and Climate Science*. Cognella Academic Publishing.

4

THE THERMOSTAT

The selective absorption of the atmosphere is … not exerted by the chief mass of the air, but in a high degree by aqueous vapour and carbonic acid. The influence of this absorption is comparatively small on the heat from the sun but must be of great importance in the transmission of rays from the Earth.

Svante Arrhenius, 1896[1]

Many of us in Australia have a rather comfortable life. If it gets a bit hot, we can set the thermostat to cooler and let the air conditioner do its job. If it is a cold winter's day in Canberra, on go the heaters and the house warms up to its automatically controlled temperature. We can change the comfort level as we choose.

The Earth also has a thermostat, and until recently the Earth was in charge of its own temperature setting. But much to the surprise of many scientists, and the disbelief of almost everyone else, over the past two centuries some of that control has been taken over by humans. At first we had no idea that was what we were doing, but gradually as we learned more about the way the Earth's thermostat works, it has become increasingly clear that we have the power to set the Earth's temperature.

Thermometer measurements prove that the global temperature has risen since the Industrial Revolution, and that it has not risen smoothly but with short-term rises and falls. These more rapid irregularities can be explained by well-known changes, in the Sun or in the ocean or in the air. However, the Earth's overall temperature rise has been attributed, by most climate scientists, to increasing levels of greenhouse gases in the atmosphere, and in this chapter we will examine that claim.

Throughout geological time, at a rate as slow as the drifting of the continents, we know from reading the story in the rocks that the world's climate has undergone major changes in response to its two dominant drivers, the Sun and atmospheric carbon dioxide (CO_2). (In Chapter 8 I explain how we know what the atmosphere was like long ago.) Although such knowledge is important in understanding climate change, there is no reason for complacency just because major changes happened in the past. Not counting water vapour, which varies regionally from 0 to about 4 per cent depending on the humidity, at present the atmosphere contains 78 per cent nitrogen gas, 21 per cent oxygen, and the last 1 per cent is made up mostly of argon with trace amounts of other gases, including almost 400 parts per million (ppm) or 0.04 per cent CO_2. Long ago, the world survived with 4000 ppm CO_2 in the atmosphere and temperatures that were 10°C higher. But that does not mean that we can survive under those conditions. At the end of one of the more recent ice ages, 130 000 years ago, as the world warmed, CO_2 rose from a low of 190 ppm to a high of 290 ppm at a rate of 1 ppm every 100 years. In 2011 the increase was 2 ppm; that is, 200 times as fast.

For the past 10 000 years, the amount of CO_2 in the atmosphere stayed almost constant at about 280 ppm. Then, in about 1750 the picture changed. From the mid–18th century the concentration of CO_2 started to rise. Measurements of air trapped in polar ice sheets allow us to track the growth in atmospheric CO_2 over the past 250 years. Figure 4.1a shows the change. Except for a decade of flatness between 1940 and 1950, the curve is continuously upward, with no more than 2 or 3 ppm variation. By 1950 the levels had reached 310 ppm, a higher concentration than at any time in the previous 2 million years. There is another graph that looks rather similar to Figure 4.1a, isn't there? The last 250 years of the Hockey Stick has a remarkably similar shape (Figure 4.1b). Could that be a coincidence?

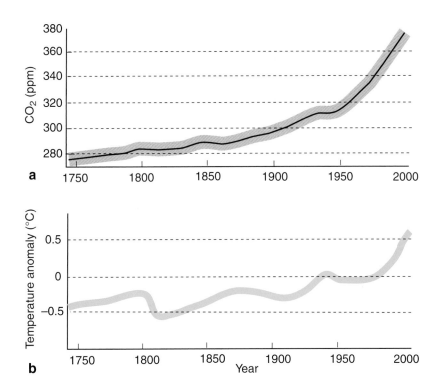

Figure 4.1 a) Atmospheric CO_2 content, based on ice core records from 1750 to 2000, combined from eight different studies. The pale grey strip covers the range of variation between the different studies. Note that the increase in CO_2 over these 250 years is 100 ppm. Source: MacFarling-Meure et al., 2006[2]. b) The last 250 years of the northern hemisphere temperature record, simplified and smoothed. Source: Mann et al., 2009.[3] Note the general similarity to the shape of the CO_2 curve of Figure 4.1.

THE KEELING CURVE

The story of contemporary measurement of CO_2 is the story of one man, Charles David Keeling. Before 1950, measurements of atmospheric

CO_2 made by sampling the air showed such wide variation in both time (day, night, summer, winter) and place (latitude, longitude, altitude) that some suggested the measurements might be useful in plotting the movement of air masses. Published values varied from 150 ppm to 350 ppm. In 1956 Keeling was doing post-doctoral research into the way CO_2 establishes equilibrium between the air and surface water. To do this he had to measure the CO_2 concentration in the air and in water, and to do that he had to design his own measuring instruments able to measure the gas to better than one part in a million. Because he enjoyed the outdoors, Keeling took his air-sampling apparatus to a variety of mountain areas and soon discovered that wherever he went, provided there was a bit of wind and turbulence, the air contained 310 ppm CO_2; it did not vary at all.

To cut a long and eventful story short (you can read his own version here[4]), Keeling was asked to set up continuous CO_2 measuring equipment in the newly established US Weather Bureau Station high on Mauna Loa in Hawaii for the proposed International Geophysical Year of 1957–58. Another instrument was taken to an Antarctic weather station. The first year of measurement showed a CO_2 variation from 310 to 315 ppm, and Keeling was worried that the results were too erratic, but after another year the results showed a clear seasonal pattern, highest at the end of the northern winter, lowest in the autumn (Figure 4.2). He had discovered that the atmospheric CO_2 content fell during the spring and summer-time growth of the vast northern hemisphere deciduous forests and grasslands as they took in CO_2, and then rose again during the winter-time hiatus in plant growth but when animals, who did not give up emitting CO_2, remained active. Keeling had seen the breathing of the Earth.

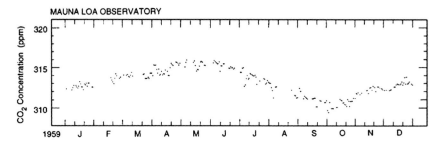

Figure 4.2 Keeling's 1959 measurements of CO_2 at Mauna Loa, Hawaii. Source: Keeling, 1998.[4]

For the following 10 years, the Mauna Loa Observatory continued to measure CO_2 and the seasonal change of 5 ppm was maintained, but there was a new feature. Each year the early spring high value and the autumn low value were a little higher, each by almost 0.5 ppm. There was a steady trend of increasing atmospheric CO_2. Twenty years earlier, Guy Callendar, an English steam technologist, had concluded that 'fuel combustion had added about 150 million tons of carbon dioxide to the air'.[5] Keeling similarly attributed the increase to the burning of fossil fuels – coal and oil.

Fifty years later, and everything Keeling discovered at Mauna Loa has been repeated around the world. The rise in atmospheric CO_2 has been inexorable, as shown in Figures 4.1 and 4.3, and as I write it stands at 396 ppm; in 1750 it was 275 ppm. Another greenhouse gas, methane (CH_4), has risen in the same way; since 1750 it has more than doubled in its atmospheric concentration.

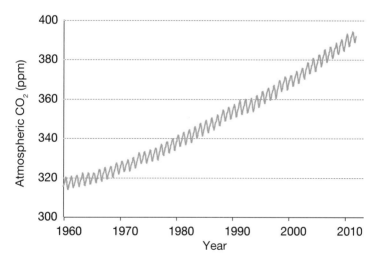

Figure 4.3 Atmospheric CO_2 measurements at the Mauna Loa Observatory measured in ppm, 1960 to 2009. This graph might look schematic but these are real data. The line shows the seasonal oscillations and the inexorable increase discovered by Keeling. Source: US Carbon Dioxide Information Analysis Center.

What is the connection between the Keeling curve, as it has come to be known, and global warming? CO_2 is a potent greenhouse gas; it is an important part of the atmosphere's blanket that keeps the world at a habitable temperature. Its amount in the atmosphere has increased by 35 per cent since pre-industrial times and that increase creates more heat absorption by the atmosphere. In 1970, the International Radiation Investigation Satellite (IRIS) measured the Earth's radiation – its heat loss. The measurement was repeated by the Tropospheric Emission Spectrometer (TES) satellite in 2006. A comparison of those measurements shows there has been a distinct decrease in the amount of heat leaving the Earth.[6] The 'blanket' is getting thicker and the setting of the thermostat is rising. The increased greenhouse effect of increased CO_2 over those 36 years is observational fact, not theory.

BALANCING THE HEAT

In Chapter 3 we saw that reflection by clouds, snow, deserts, forests and the ocean removes 30 per cent of the Sun's heat. Other than shading by volcanic or industrial aerosols, that leaves 70 per cent to warm the atmosphere and surface. If that were all, the Earth would just get hotter and hotter until the oceans boiled, but of course it does not. Since it is at a temperature above the cold of space (about minus 270°C), the Earth is itself a radiator. The second part of the heat balance is the Earth's own out-bound radiant energy, which should balance the 70 per cent left of the incoming heat. It eventually has to balance; if the Sun gets hotter, so will the Earth and it will radiate more heat until it reaches a new balance temperature. If the Sun's irradiance is reduced, the Earth cools down and radiates less heat away, again until a balance is struck. This is what made the ice ages.

But what happens if the change is from the Earth's side of things rather than the Sun's? Suppose through some biological disaster the Earth loses all its forests and grasslands and they become deserts? The amount of reflected heat will increase and the Earth will cool down. That might make more snow, which would reflect more heat and cool the Earth even more. Conceivably, if nothing happened to stop the trend we would end up with a frozen planet. Geologists suggest there was a period earlier in the Earth's history when it was frozen over, referred to as 'snowball Earth'. This was postulated to have happened about 700 million years ago, and the process leading to it was just as I described (though not the result of loss of plants, since there weren't any). The problem with the theory was understanding how the Earth escaped from this freezing, because it certainly did not last forever. The suggested answer lies in what are called greenhouse gases: gases that absorb heat.

The two most important absorbers of heat in the atmosphere are water vapour and CO_2. By stopping some of the heat from escaping the Earth, these two were largely responsible for keeping the Earth's surface temperature at the average of 14°C rather than the moon's average of minus 23°C. They are our thermostat. That they affect global temperature has been known for 150 years. In 1861, John Tyndall discovered that water vapour and CO_2 were able to absorb heat. In 1896, Svante Arrhenius concluded that doubling atmospheric CO_2 would increase global temperatures by 5°C, and he speculated that the ice ages might have been caused by a lower content of atmospheric CO_2. As long as the amount of these gases remains stable, the thermostat remains fixed and there will be no global warming, no climate change. All it takes to move the thermostat is to add or subtract some greenhouse gas. And that is exactly what we have done.

For the past 10 000 years, the atmosphere has held a constant amount of CO_2, round about 280 ppm. True, that is not much, but if it were prussic acid (hydrogen cyanide) that amount would kill you immediately. On the other hand, if it was not for that 280 ppm CO_2 there would be no life on Earth. All the carbon in your body ultimately came out of the air. More importantly to the topic of climate change, this 280 ppm of CO_2 helped to keep the thermostat at 14°C. It is now estimated that doubling the amount of CO_2 would raise the global temperature by at least 3°C.

Carbon dioxide, water vapour, methane and some others are collectively referred to as greenhouse gases because they seem to work like a greenhouse does. It is a pity that such a commonplace comparison as a greenhouse is wrong.

To see where the comparison fails, think about how you warm up when you are cold. You might stand in front of a fire and let the heat *radiated* by the fire warm you, or you could get into a hot bath so the heat in the water is *conducted* into your body. The third method of heat transfer, known as *convection*, involves warm air or water rising and cooler air or water sinking, which is how a pot of water gets hot all through, even though the heat source is at the bottom. Convection is the way the air moves heat around the world.

Only one of these three ways of moving heat around can work for the Earth in space. The Earth cannot conduct its heat away because there is nothing out there to carry the heat. It cannot lose heat by convection because the atmosphere stops about 100 kilometres up. All that is left is radiation. Radiant heat has a range of energy, its own spectrum, and this energy is sent out as waves. The waves are tiny; 100 waves might occupy only one millimetre, and their energy is related to their wavelength, the distance between one wave crest and the next. If you have ever waited in the surf for a good wave you will know what wavelength is – it is the distance between waves. Heat waves are very close together and move a lot faster than water waves – 300 000 kilometres a second – and the shorter their wavelength the hotter they are. When you switch on a bar heater, almost immediately you will feel a little warmth radiating from it. This is low-energy heat and for heat it has a long wavelength, about a thousandth of a millimetre; that is, one micrometre (Figure 4.4). After a while the bar glows a dull red; it is now radiating higher-energy heat and also some light. Eventually it becomes orange-red in colour; it is radiating even higher energy heat as well as the light you can see, with a shortest wavelength of about half a micrometre. It is still radiating the lower energy, longer wavelength heat as well as all the wavelengths of its spectrum in between.

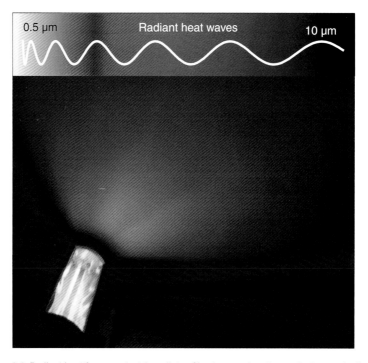

Figure 4.4 Radiant heat from an electric radiator. Shorter wavelengths are hotter. μm is short for micrometre, a thousandth of a millimetre.

GARDEN GREENHOUSE

A greenhouse works because it lets in the light and heat from the Sun, which warms the plants and the air in the greenhouse. The glass walls and roof prevent the warm air from flowing out by *convection* – warm air rises but the roof stops it escaping. Glass is not a good conductor of heat, so the warmth of the air in the glass house is *conducted* away slowly. The greenhouse does lose heat by *radiation*, but enough heat is trapped inside by the glass for the greenhouse to remain warmer than the rest of the garden until sunset (Figure 4.5).

Figure 4.5 How a greenhouse works. High-energy heat from the Sun warms the interior, while lower-energy heat radiates out, keeping the greenhouse from overheating. Some heat escapes if a window is opened and some is conducted through the glass to the outside air.

THE EARTH'S GREENHOUSE

The Earth does not have a glass case around it. Greenhouse gases keep in the Earth's heat quite differently from the way the glass of a greenhouse does; Figure 4.6 depicts what happens. The Earth is being heated by the sun, which makes the Earth itself a radiator. Not a very hot radiator, but a heat radiator nonetheless. The Earth radiates heat of low energy; that is, of longer wavelengths than a bar radiator. The Earth's radiant heat, Earthshine, heads upward towards space and has to pass through the atmosphere to get out. Most Earthshine escapes; if it did not we would boil. But some parts of the heat's wavelength spectrum are absorbed by water vapour and carbon dioxide, and this is called the 'greenhouse effect'. Water vapour absorbs some of Earthshine's high-energy heat and also some of the low energy. A narrow, central part of the Earth's heat spectrum is absorbed by carbon dioxide and another part by methane. The absorbed heat stays with us and keeps the Earth at about 14°C; the rest is lost.

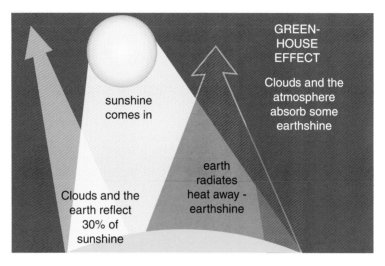

Figure 4.6 The Earth's heat budget. Incoming heat from the Sun is somewhat reduced due to direct reflection by clouds and the Earth's surface. Over 70 per cent of the radiation from the warm Earth, Earthshine, passes through the atmosphere and out into space. Heat is kept in a little through absorption of Earthshine by clouds, but predominantly by the so-called greenhouse effect: the absorption of heat, mainly by water vapour, CO_2 and methane in the atmosphere.

Figure 4.7 shows how, on a cloudy day, the Earth's radiation gets cut out by water vapour, CO_2 and ozone before it reaches outer space. The graph looks superficially like Figure 3.2, but the big difference is that Figure 4.7 starts where Figure 3.2 ends. Most of the sun's radiation tails away at longer wavelengths than 2 micrometres (2000 nanometres in Figure 3.2), whereas the Earth's outgoing heat radiation only begins at around 2.5 micrometres. In Figure 4.7, the smooth curve is the radiation that leaves the surface and the jagged curve is what actually gets through a cloudy atmosphere. About 25 per cent more heat escapes on a clear day (or night). Water vapour absorbs heat in several different parts of the radiation spectrum, amounting to about 60 per cent of all the absorbed heat. Carbon dioxide absorbs an appreciable amount round about the middle of the Earth's heat-radiation spectrum.

There are instances, albeit rare, of physicists who have misunderstood the effect of increasing CO_2 on the atmosphere's ability to absorb heat. In 2007 a Hungarian scientist, Dr Ferenc Miskolczi, published

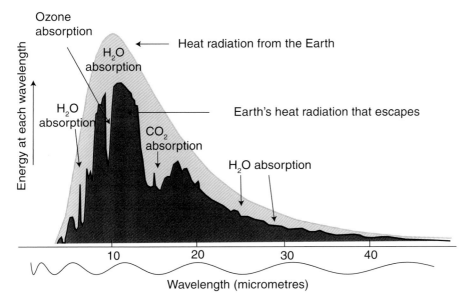

Energy at each wavelength

Ozone absorption

H_2O absorption

Heat radiation from the Earth

H_2O absorption

Earth's heat radiation that escapes

CO_2 absorption

H_2O absorption

10 20 30 40

Wavelength (micrometres)

Figure 4.7 Spectrum of Earthshine. The upper curve represents all the heat radiated from the surface of the Earth. Red represents the heat that gets through the atmosphere on a cloudy day and escapes into space. The difference between the heat radiated and the heat that is absorbed is the pink part, the remaining heat that keeps the Earth warm. Notice how big a bite CO_2 absorption takes out of Earthshine. Source: after Kiehl & Trenberth, 1997.[7]

a peer-reviewed paper postulating that, as CO_2 levels increase, atmospheric water vapour decreases in amount to compensate, so that the total atmospheric absorption of the Earth's outgoing heat remains constant.[8] His concept is based on the idea of energy balance – roughly, 'what comes in must go out; the temperature therefore stays the same as long as the Sun stays the same'.

Miskolczi's paper has been refuted several times, mostly in internet postings, which say that it is meaningless since it is based on the incorrect application of three laws of physics. Thomas Huxley's famous saying about 'the slaying of a beautiful hypothesis by an ugly fact' finds another application here. Observations show that the water vapour content of the atmosphere is increasing in response to global warming, not reducing, as Miskolczi's theory predicts.[9]

The idea that increasing the level of atmospheric CO_2 has no effect on heat absorption is also the theme of a paper by Dr Heinz Hug.[10] Hug's conclusions stem from the observation that when a beam of infra-red radiation of 15-μm wavelength is shone through a long tube containing air, the heat beam is completely absorbed. Therefore, it is deduced, the CO_2 in the air already absorbs all the radiant heat from the Earth it can or ever will; adding CO_2 to the air has no effect. From this conclusion it is claimed that all 15-μm wavelength heat radiated from the Earth is absorbed within 10 metres of the Earth's surface.

The physics of infra-red radiation from the Earth is a lot more complicated than is described in the experiment, involving multiple emissions and absorptions of heat all the way up to the outer atmosphere. Infra-red of wavelength 15 μm is radiated throughout the atmosphere when higher-energy heat is absorbed by other molecules and then re-radiated. But there is no need to theorise about this. Satellites have been used to measure the heat radiated from the Earth. When measurements are compared, both CO_2 and methane are shown to be absorbing more heat now than they were in 1970.[11] The evidence says that CO_2 is NOT absorbing as much as it ever can, but rather that the amount of heat lost to space has declined between 1970 and 2006 because the amounts of CO_2 and methane in the atmosphere have increased. Furthermore, high levels of CO_2 in the past were coupled with warmer climates (see Chapter 8). The global temperature has changed as the atmospheric CO_2 levels have changed, therefore the atmosphere's ability for CO_2 to absorb heat is not saturated at current levels.

The composition of the atmosphere varies quite a bit because of local and seasonal factors. Nearby freeways or factories might increase the CO_2 content of the air. Nearby cane-fields or forests may remove some CO_2 locally by photosynthesis. To avoid this sort of variation, scientists choose to sample the atmosphere in places where they can be sure that most of the time there will not be such problems. One of the main laboratories for CO_2 measurements is high on Mauna Loa, Hawaii, where there are no nearby forests, cars or factories. There is a nearby volcano that emits some CO_2, so occasionally the Mauna Loa data are affected, but this is not something the scientists do not know about. Similar laboratories in the Arctic and Antarctic and in other 'clean' environments, such as Cape Grim in Tasmania, all yield the same steady increase in atmospheric CO_2. Of all the measurements related to climate change, the knowledge of atmospheric CO_2 is probably the least open to question.

CLOUDS AND WATER VAPOUR

Clouds reduce the Sun's radiation. Earlier we saw that clouds reflect sunlight back into space, and in doing so they act to cool the Earth. However, clouds are made of water vapour, which is a greenhouse gas, and they absorb the heat radiated from the surface and help to warm the Earth. It has been thought that clouds have an overall cooling effect,[12] though researchers such as Andrew Dessler at Texas A&M University have concluded that clouds act as a feedback mechanism that amplifies temperature change.[13] Understanding the way clouds affect local and global temperature and climate is proving to be one of the biggest challenges facing climate scientists.[14]

Clouds are not just water vapour; they are masses of tiny water droplets, so tiny they do not fall out of the sky. Clouds are the first visible sign that the water vapour in the air is condensing to water droplets. Have you ever tried to photograph a beautiful mountain peak, waiting for the cloud floating over it to pass, only to realise it never will? The wind is pushing the air and its load of water vapour up the mountainside towards the peak. As the air rises it cools, and when it gets high enough the water starts to condense into droplets and makes a cloud. Exactly the same thing happens when moisture forms on the outside of a glass of cold drink. Once past the peak the moist air falls, warms up a little and the droplets vanish back into invisible vapour, and you can never get a clear photo of the peak even though the sky everywhere else is cloudless.

The cloud that spoilt your photo started off as water in the ocean. The Sun's heat warmed the surface – remember those shallow pools on the beach that are warm even when the sea itself is cold? Warmer water evaporates more readily than cold water. The warmest water turns to vapour and the water left behind gets cooler. The air is warmer than the ocean, so the water vapour takes up heat from the air and is moved by the wind until it reaches a cooler place, where it condenses and falls as rain or snow. This is one way heat is constantly exchanged and cycled between the atmosphere and ocean.

The amount of water vapour in the air varies considerably; we feel it as humidity. In the desert and at the poles the humidity is low, almost zero at times. In the tropics, during the wet season the humidity can be stifling, and when it reaches 100 per cent it rains. A humidity of 100 per cent means there is as much water vapour in the air as can stay there without condensing into droplets; it does not mean the air is 100 per cent water. In Darwin, at a temperature of 29°C and a humidity of 83 per cent, the air contains 3.4 per cent water in a given volume. On a cold day in Antarctica, at the same relative humidity the air would only contain 0.04 per cent water.

Evidently, the warmer the air the more water it can hold. And the more water it holds the more it absorbs the heat radiating from the Earth's surface. On the basis of chemical theory, the temperature rise since 1970 of 0.5°C should cause a 4 per cent increase in the amount of water vapour in the atmosphere. Using satellite data and earlier measurements from balloons, climate scientists have found that the amount of water vapour over the oceans has increased by 4 per cent since 1970.[15] When the theories and observations of science fit together so closely they strengthen our confidence that we understand at least some of the impact of global warming on other aspects of the climate.

This increase is very important in terms of the greenhouse phenomenon. Water vapour is the most effective greenhouse gas (Figure 4.8, p. 66), responsible for 60 per cent of the total heat absorption by the atmosphere. Increasing the global temperature increases evaporation from the ocean and so increases the total water vapour. Here, we have a feedback loop – a critical feedback loop. The temperature rises, more water evaporates into the air, this traps more heat and so the temperature rises some more. No matter what caused the initial rise, it is amplified by the consequent increase in water vapour. Any increase in water vapour causes an increase in the global temperature.

GREENHOUSE GASES

Leaving aside water vapour for the moment, as well as the nitrogen, oxygen and argon (totalling 99.9 per cent of the atmosphere), the 'greenhouse gases' are left: mainly 390 ppm CO_2, 1.7 ppm methane (CH_4) and 0.3 ppm nitrous oxide (N_2O). Despite their small amounts, all of these absorbing gases help to keep the Earth warm.

The effects of water vapour and CO_2 are commonly misunderstood. The atmosphere has more than 100 times less CO_2 than water. How, it is asked, could even doubling the minuscule amount of CO_2 have any significant effect compared with the huge effect all that water vapour must have? Look again at Figure 4.7. You can see the size of the bite that 'minuscule' amount of CO_2 takes out of the Earth's radiation spectrum; that is, how much it absorbs. You can also see that there is room to bite out some more if the CO_2 content rises. The water vapour content can rise if it likes, it just does not change the fact that CO_2 is a potent greenhouse gas with its own effects.

On the issue of rising CO_2 and the resulting trapping of the Earth's heat, the laws of physics do give us a break. Doubling the amount of

CO_2 or water vapour does not double the heat absorption; fortunately, it increases it by only about 30 per cent. This is a consequence of what is known as Beer's law, which states if you increase the amount of an absorbing gas by X times, the amount of heat it absorbs will increase by $\log(X)$ times. Since $\log(2) = 0.301$, doubling the CO_2 only increases the absorption by 30 per cent.

FORCING

Forcing or, more explicitly, 'radiative forcing', is a term climate scientists use as a simple measure to quantify and rank the many different influences on climate change in terms of energy balance. Thus, forcing refers to the extent to which any agent of change, be it from the Sun, in the atmosphere, in the ocean or on the ground, will force a change in the previous balance that dictated the climate. To be specific, it is the energy change caused by that agency per unit area of the globe as measured at the top of the atmosphere.

Imagine you are sitting on the verandah on a warm summer's day and that you are at a very comfortable temperature. If you then put on a jumper, it will force you to get hot. Putting on the jumper is a positive forcing. If instead you spray yourself with a water mist, the water will force you to feel cooler. The water has a negative forcing.

Forcings occur continuously, from day to night, as the clouds come and go, or as plant cover changes with the seasons, thereby changing the amount of the Sun's heat that is reflected or by the natural cycle of CO_2 variation between seasons. The Intergovernmental Panel on Climate Change evaluates forcings in terms of their impact since 1750. It is necessary to choose a starting time, because forcing is defined in terms of *change* from a previous situation, so some prior condition must be specified. There is an equation that defines forcing; for CO_2 the change in forcing is 5.85 times the natural logarithm of the ratio of the CO_2 concentration before and after. Taking 1750 as the initial time, when CO_2 stood at 280 ppm, the change in forcing (symbolised ΔF) from then to the 2005 CO_2 level of 385 ppm is:

$\Delta F = 5.85\ln(385/280) = 1.66$ watts per square metre.

Other values are:[16]

Methane	+0.48 watts per square metre
Halocarbons	+0.34 watts per square metre
Ozone	+0.3 watts per square metre
N_2O	+0.16 watts per square metre
Albedo	−0.2 watts per square metre
Aerosols	−1.2 watts per square metre

(See also the box in Chapter 12 on climate sensitivity – the increase in temperature that would be caused by doubling atmospheric CO_2 from its pre-Industrial Revolution level.)

Water vapour and CO_2 are not the only absorbers; methane (CH_4) is another (Figure 4.8). Per unit mass, methane is a much stronger absorber of infra-red than CO_2, about 30 times as strong; however, it is much less abundant, only about 1.8 ppm. Even so, that gives methane the equivalent absorption ability of another 50 or so ppm CO_2, so it is not inconsequential. Ozone is a third greenhouse gas, but more important to us because it blocks incoming ultraviolet rays and a fourth is nitrous oxide (N_2O).

Figure 4.8 The contribution of various greenhouse gases to heat absorption on a clear day. Without these gases in the atmosphere, Earth's average temperature would be well below freezing; about minus 23°C, not 14°C.

As a result of the absorption of these greenhouse gases, the atmosphere itself is warmed. If there were no infra-red absorbing gases, the air would let all the Earth's heat radiate back to space without hindrance, just as it lets all the light from the Sun in. If the air were infra-red transparent, the Earth's surface would be at or below freezing all over. But these gases do absorb heat radiated up from the surface, and then they radiate it out at lower energy in all directions so the air becomes warmer.

At this stage we need to digress. Atmospheric gases, like those in the Sun's and Earth's radiation, have their own mechanism for balance. A gas such as CO_2 had its origin in a volcano. Volcanoes have been erupting since the world began, and at present they put about 300 million tonnes of CO_2 into the atmosphere annually, which sounds like a lot until you realise that the total amount of CO_2 in the atmosphere is 3 trillion tonnes. The atmosphere therefore has 10 000 times more CO_2 than the annual input from volcanoes. Looked at in this way, it would take only 10 000 years for volcanoes to provide the current amount of CO_2. Volcanoes have been spouting CO_2 for 4500 million years, so where is all the rest of it? There must be a graveyard somewhere where CO_2 goes to be buried.

You need not look very far. Put some lime on the garden and you have just used a bag of CO_2 burial ground in the form of calcium carbonate ($CaCO_3$). That $CaCO_3$ was probably mined from a limestone quarry, and limestone is basically the accumulated bodies of seashells and other kinds of sea creatures. What happens is that the CO_2 in the atmosphere gets into the ocean. The oceans contain 120 trillion tonnes of CO_2, or about 40 times as much as is in the air. In the ocean, fish, oysters, clams, corals and micro-organisms extract the CO_2 from the water to construct their bones or shells, and when they die their bodies fall to the bottom, are covered over by more shells and sand, and eventually become rock. End of the CO_2: it has gone out of the volcano and into the air, from the air into the sea, from the sea into a shellfish and from it into a rock. So the volcano can spew out as much CO_2 as it likes; all the CO_2 will eventually be converted to rock. If there is not enough CO_2 to maintain all those creatures, some of them die young and so do not reproduce, and their numbers decline until there is a new balance between CO_2 dissolving in the sea and CO_2 being converted to rock.

Now, all that processing does not happen overnight. It is slow; it is the Earth's long-term balance between CO_2 produced and CO_2 buried. The short-term process is also very well known; it is the balance between the consumption of CO_2 and water by plants that turn it into sugar and then into starch and other carbohydrates, and the consumption of plants by the animals who burn the plants – that is to say, oxidise them, to gain their energy and in doing so convert the carbohydrates back to CO_2 and water.

To be fair, this is a simplification of a great many activities involving CO_2 that are occurring in the air, soil and water. Oxygen is obviously important, as are the other essential elements of life, phosphorus, potassium, nitrogen and sodium, and many of the rest of the 90 elements in the Earth's crust. What is important in the consideration of the movement of CO_2 between Earth and air is that it was pretty much in balance. In the natural scheme of things, the CO_2 taken up by plants is returned some time later, either when they die and decay, or when the plants are eaten by animals. Animal waste products include CO_2.

MORE ON FEEDBACK

Getting back to the ocean, we saw that a lot of CO_2 is dissolved in it. It is possible to dissolve a lot more sugar in hot water than in cold, but the opposite is true when CO_2 dissolves in water; the hotter the water the less CO_2 it can dissolve. If all the world's oceans were to warm by 1°C, estimates suggest it would release enough CO_2 to increase the atmospheric concentration by about 30–40 ppm.[17] You can probably already imagine what is going to happen if the ocean temperature rises. Some CO_2 will be released and will add to what is already blanketing the Earth, and that will trap more of the Earth's radiance and so the temperature will rise some more. This is a second feedback loop; a small increase in temperature is amplified both by increased water vapour and by the release of more CO_2 from the oceans.

There is a third type of feedback (see Figure 4.9, p. 69). If the atmosphere warms, some of the polar ice will melt. Ice has a very high reflectance; melting some of it reduces the amount of the Sun's heat that is reflected back into space, which means more is absorbed, which means the global temperature will increase. As an ice age wanes, for example, the snow fields at the edge of the continental glaciers are replaced by tundra, where plants can grow, grateful for the warmer climate and the increasing CO_2 levels. Ever more plants, ever more trees, but not ever more CO_2 because it is taken up by the plants, soil organisms and ocean creatures as it is released by the oceans. Provided the CO_2 increase is slow, the feedback will be slow and the biological processes that take up the CO_2 can keep pace and constantly maintain an equilibrium. And while there is almost no theoretical limit to the amount of warming a greenhouse gas can do, because of Beer's law, increasing the atmospheric CO_2 or water vapour 10-fold only doubles its greenhouse effect. The accelerated temperature rise, caused by the feedback mechanisms of increased oceanic CO_2 release and water vapour, gradually slows because of this.

We saw in Chapter 3 that the start of each of the recent ice ages was a response to changes in the distribution of the Sun's radiation reaching the Earth; not, for example, a response to declining CO_2 in the atmosphere. The change in radiation was small but it was amplified. The ocean cooled a little and took more CO_2 out of the air. The cooler atmosphere

Figure 4.9 Three feedback loops that enhance global warming. If the temperature rises, the melting of snow and ice reduces the reflection of the Sun's heat back into space. At the same time, more water evaporates and more CO_2 bubbles out of the oceans, increasing the greenhouse effect, so the temperature rises some more. Source: Arctic images modified from those published by the US National Snow and Ice Data Center.

became a little drier, so the water vapour greenhouse effect began to reverse. Summers then were cooler and less snow melted, more sunlight was reflected and slowly the climate became glacial. Slow and steady allows everything and everybody to adjust. Fast and furious does not.

SUMMARY

Topic	Observation	Key statistic	Conclusion
The atmosphere	Acts as the planet's thermostat.	Maintains global temperatures well above −23°C of the moon.	Atmospheric CO_2 rise has upset the Earth's thermostat.
Atmospheric CO_2	Slowly varies over geological time. Traps the Sun's heat, helping to keeping the Earth from freezing.	The current rate of increase of CO_2 in the atmosphere is 200 times quicker than in previous centuries.	Rapid increase of CO_2 in the atmosphere since 1750. Increase has caused global warming.

Topic	Observation	Key statistic	Conclusion
Atmospheric CO_2 *(cont'd)*		Responsible for almost 30% of the greenhouse effect.	
The Keeling curve	Research enabling accurate measurement of atmospheric CO_2	CO_2 in the atmosphere was 396 ppm in 2012, compared to 280 ppm in 1750.	Atmospheric CO_2 steadily increasing.
Water vapour	Water vapour in the atmosphere traps the Sun's heat, also helping to keep the Earth from freezing.	Responsible for 60% of the greenhouse effect.	Increasing water vapour as the world warms creates a feedback loop, increasing the temperature.
Clouds	Reflect the sun and also absorb the sun's heat.	Net effect uncertain.	More research is needed.

The answers to the first two questions are now becoming clearer.

1) Is the climate changing and 2) what can change the climate? Answers to question 2 include changes to the Sun, changes in the Earth's orbit, changes in Earth surface reflection of the Sun's radiation, changes in concentrations of greenhouse gases, changes in ocean temperature and changes in ocean currents. The globe is warming and the reason seems to be because there has been an increase in the amount of greenhouse gases in the atmosphere since 1750. The additional CO_2 and methane block enough heat from leaving the Earth to cause a rise in the global temperature. But is that enough to change anything else? Are there changes in ocean and atmospheric behaviour? Is there really *climate* change? In the next three chapters we will look at other aspects of climate: rainfall, the world's ice fields and the ocean.

FURTHER READING

Climate Change 2007 – The Physical Science Basis Contribution of Working Group I to the Fourth Assessment Report of the IPCC (ISBN 978 0521 88009–1 hardback; 978 0521 70596–7 paperback).

Pittock AB (2009) *Climate Change: The Science, Impacts and Solutions.* CSIRO Publishing.

5

DROUGHTS AND FLOODING RAINS

In the six hundredth year of Noah's life … the windows of heaven were opened.
And the rain was upon the Earth forty days and forty nights.

Genesis 7: 11–12, *The Holy Bible*, Revised 1884

RAIN

'We were three days short of a Biblical record,' said Foreign Minister Habib Bourguiba Jr. He was not smiling. For 38 days in September and October [1969], *rain fell steadily on Tunisia, leaving 600 people dead, destroying 70 000 homes, and making refugees of 300 000 of the nation's 4 500 000 people.*[1]

The rain started with a deluge of 400 millimetres in 24 hours, as much as the mean annual rainfall for the region in one day. In some places, up to a metre of topsoil was washed into the Mediterranean Sea. Bridges built in the time of the Romans were destroyed, making this possibly the most intense rain event in that country in 2000 years. A climactic event no doubt, but did it signal climate change?

Plants know all about climate. In the long run, the weather is not so important to them. Certainly, many plants die during a drought, but if they grow in drought-prone regions they will have evolved strategies to allow their seeds to survive and reproduce when the rains do come. New settlers in a country are not so well adapted. By 1860, most of the arable land near Adelaide, the capital of South Australia, had been taken up. The rainfall had been good for the previous 20 years, and this encouraged settlers to move north and try their luck at farming the country between Port Augusta and the Flinders Ranges. But then came a five-year drought, and many of the northern farms failed. The Surveyor General, George Goyder, was sent out from Adelaide to determine where farming could be profitably pursued; that is, whereabouts farmers could rely on adequate rain. There were no rainfall records in 1860 but Goyder did not need them. He reasoned that the plants knew what the climate was like,

Figure 5.1 Bluebush meets eucalypt on South Australia's Eyre Peninsula. Goyder used the southern limit of bluebush-dominated country to mark the northern limit of farming.

A SHORT INTRODUCTION TO CLIMATE CHANGE

and he drew a line on a map marking the southern limit of saltbush and bluebush.[2] Goyder's Line then marked the northern limit of farming (see Figure 5.1). It was a vegetation line – a climate line – not a weather line.

The agricultural city of Wagga Wagga in New South Wales was in drought from 2000 to 2009. The graph of average annual rainfall shown in Figure 5.2a (in this graph each column is the average of that year and the four preceding years) shows an alarming decline from over 700 millimetres in 1996 to not much over 400 millimetres in 2009. Fifteen years of decline; surely good evidence for climate change? Not necessarily. Certainly, the data show the drought in that part of the world at the start of the 21st century, but the time frame is too short to declare it as evidence of climate change. Figure 5.2b shows the full record for Wagga Wagga, going back to 1890. Over those 120 years there have been five other periods of steadily declining average rainfall, shown by the arrows.

The longest stretch of low rainfall for Wagga Wagga was from 1895 to 1915, during which period there were only three years above the average

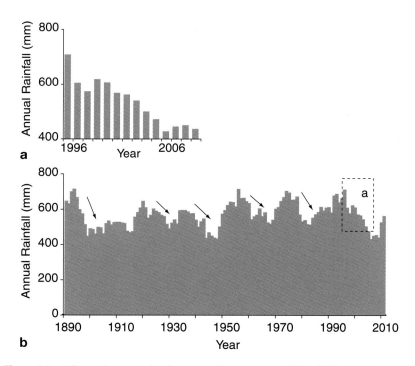

Figure 5.2 a) Wagga Wagga rainfall, five-year rolling average, 1996 to 2009. b) Full record of Wagga Wagga rainfall, again five-year rolling average. Periods of steadily declining rainfall are shown by arrows, the dashed-line box is the section shown in a). Source: Australian Bureau of Meteorology

rainfall of 560 millimetres. On the basis of the rainfall data from this meteorological station, no positive conclusion about climate change over the past 120 years could be made. What the Wagga Wagga data show is the variability in Australia's weather. If you live on Australia's south-western coastal region, say in the village of Wilgarup in Western Australia, you might well appreciate that the climate has changed. Here, the rainfall has steadily declined from over 1000 millimetres in the 1920s to a meagre 750 millimetres in 2000 (Figure 5.3a). By contrast, in northern Australia the opposite is true. At Victoria River Station in the Northern Territory, rainfall has risen from about 500 millimetres to over 850 millimetres in 100 years (Figure 5.3b). Because these are individual records, they also do not show proof of climate change, but they are certainly suggestive.

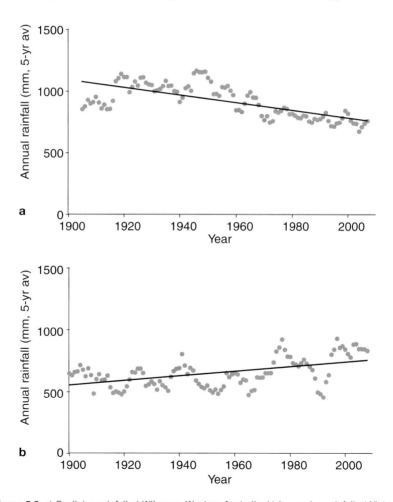

Figure 5.3 a) Declining rainfall at Wilgarup, Western Australia. b) Increasing rainfall at Victoria River Downs, Northern Territory

A SHORT INTRODUCTION TO CLIMATE CHANGE

Drawing conclusions about any national climate change needs both long-period records and a wide coverage across Australia, and this is what the Australian Bureau of Meteorology provides in the form of annual summaries of rainfall for the entire continent and for various regions of Australia covering the past 100 years.[3]

Unlike the temperature, which keeps on rising, there is no single direction of change in rainfall. Contour maps of rainfall trend provided by the Bureau of Meteorology and shown in Figure 5.4 (p. 76) show that the long-term trends in rainfall are different in different parts of the continent. Since 1970, all of eastern Australia and the far south-western coast have become particularly drier whereas the west and particularly the north-west have become appreciably wetter.

One immediate result of reduced rainfall is reduced river flow. The Murray River flows through much of Australia's most productive farm-land and is heavily used for irrigation. The 100-year average of water flowing into the Murray is a little under 10 gigalitres a year, falling to half that amount during drought. Inflow has steadily declined since 1992. By 2009, the Murray was getting less than half the historic worst inflow and only one-fifth of the overall average inflow. In south-western Western Australia, the same kind of trend is evident in the inflow to Perth's dams as it is also around Sydney and Brisbane. These rivers are all in areas where the rainfall maps show evidence for a decrease in rainfall over the past 40 years, and the evidence does seem to point more towards climate change than variable weather.

BEYOND AUSTRALIA

If we turn now to the Earth, the kind of variability in rainfall that Australia has seen is repeated again and again. If Tunisians were hoping that the floods of 1969 were an indication, albeit rather too violent, of more rains in the future, they must be disappointed, for the rainfall across northern Africa has not changed since then. But changes have taken place to the south of the Sahara in a regional band known as the Sahel; it includes the countries of Niger, Chad, Senegal, Sudan, Mali and several smaller countries. Figure 5.5 (p. 77) shows the land on the road north to Timbuktu in Mali. Across the Sahel, the climate of the past 50 years has been nothing short of disastrous. Rainfall, never abundant and once averaging between 100 and 400 millimetres annually, in some parts declined by 40 per cent between 1950 and 1980.

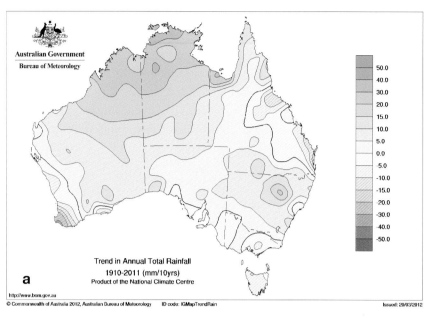

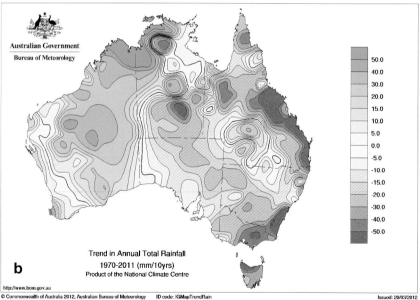

Figure 5.4 Rainfall changes in Australia. Darker green shadings show increases up to 50 mm/year and darker brown shadings show decreases up to 50 mm/year: a) 1910 to 2011, b) 1970 to 2011. Source: Australian Bureau of Meteorology, products of the National Climate Centre.

Figure 5.5 Near Timbuktu in the northern Sahel, Mali. The annual rainfall here has declined from 225 to 175 mm since 1950. Photo: Roger Hambly

Figure 5.6 shows the changes in the wet-season rainfall over the 50 years from 1950 to 2000.[4] As the rainfall has dwindled, so has the agricultural productivity of the Sahel, falling 1 per cent every year from 1970 to 2000. And, in parallel, desertification has increased. Although some improved farming practices have started to reverse the declining food

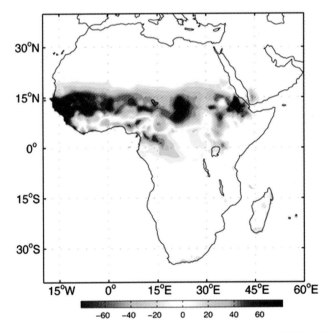

Figure 5.6 Wet-season rainfall changes in millimetres across Africa from 1950 to 2000. Source: Held et al., 2005,[4] Figure 2 (left).

production of the late 20th century, the region is still highly susceptible to drought. By early 2010, failure of the rains in the previous year had led to a 30 per cent drop in cereal production in Chad. Neighbouring Niger had its worst crops in two decades. According to the United Nations Children's Fund (UNICEF), in the Sahel 300 000 children under age five die each year from malnutrition. The drying of the Sahel is attributed to an increased Atlantic sea-surface temperature difference north and south of the equator, as well as warming of the Indian Ocean.[4]

> On the eve of the Olympics, the big story in Vancouver is still the weather. These days, a perpetual rainy mist hangs over Cypress Mountain … After a historically warm January, the mountainside is mottled with the earthen spots of a receding winter.[5]

The Winter Olympics of 2010 ran on schedule despite the lack of winter snow. It was not that there was a shortage of water in one form or another. In British Columbia over the past century, the number of days a year when it rained or snowed has increased by 50 and across Canada on average from 120 to 150. That has amounted to a 22 per cent increase in precipitation and an increased incidence of land slides, but with it has gone a reduction in snowfall.[6] Climate science predicted both of these changes: warmer temperatures reduce the amount of snow and increase atmospheric water vapour, and Canada is one part of the world where this has led to increased rainfall.

Further north, precipitation in the Yukon River Basin has increased by 28 millimetres a year since 1977, and this, possibly coupled with a small temperature rise (0.3°C) has led to an increase in discharge in the river by 8 per cent, equal in amount to the rainfall increase.[7] Such changes have surprising 'knock-on' effects; for example, increased water flow in a river stream can scour the stream bed and destroy salmon eggs.

River flows in other countries show variable trends over the past century. The Yellowstone River in Wyoming, United States, increased its flow during the 20th century from its 1780–2000 average, despite a long drought in the 1930s. The Mississippi River outflow has been higher since 1970 than it was from 1930 to 1970, but no higher than it was between 1900 and 1930. An analysis of the flow of six large rivers draining into the Arctic Ocean shows an increase of 7 per cent from 1936 to 1999.

Figure 5.7 shows the global changes in rainfall from 1951 to 2000; its significance is simply that it shows change. The Americas have seen more

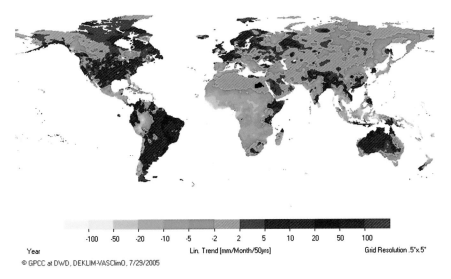

Year   Lin. Trend [mm/Month/50yrs]   Grid Resolution .5°x.5°

© GPCC at DWD, DEKLIM-VASClimO, 7/29/2005

Figure 5.7 Changes in rainfall across the globe from 1951 to 2000; purple and blue shaded areas were wetter, green areas were drier. Source: Global Precipitation Climatology Centre.[8]

rain, while Africa and much of Asia have seen less. The rains of the Indian monsoon have decreased by 5–8 per cent from 1950 to 2002.[9] Some analyses show a growing increase in the intensity of extreme rain events, such as possibly the record rainfall event over Queensland, Australia, in late February 2010. There are two ways in which rainfall change might be related to temperature change. Firstly, warmer air can hold more moisture than cooler air, which is why dew forms during the night; as the air cools some of the moisture it holds falls out as dew. Secondly, higher temperatures over the ocean increase evaporation and so put more water into the air. Together with changes in air flow, such as happen between *El Niño* and *La Niña* events, temperature rise certainly has the potential to change rainfall patterns. On the evidence of the rain, there has been climate change over the past 50 years.

STORMS

The 2005 hurricane season in North America shattered all previous records. The best known, Katrina, did more damage than any previous hurricane, while Wilma was the most intense ever recorded. In 2006, the equally intense cyclone Larry struck northern Queensland, causing widespread destruction (Figure 5.8, p. 80). Towards the end of 2009, a series of typhoons battered the Philippines. These media-attracting storms made many wonder if climate change might be responsible. Meteorologists recognise that warmer tropical waters provide more energy for cyclones and hurricanes, and it was natural that such intense and damaging storms would generate debate over global warming.

Figure 5.8 a) Cyclone Larry viewed from space by the MODIS satellite. Source: NASA image by Jeff Schmaltz, MODIS Rapid Response Team, Goddard Space Flight Center. b) Damage to a papaya orchard by cyclone Larry. Source: Geoscience Australia.

Just to clarify the words, meteorologists refer to a 'rotating atmospheric low pressure' of any intensity as a cyclone. A high-pressure event is an anti-cyclone. In North America, intense cyclones are called hurricanes; in Asia they are typhoons, while in Australia they are just known as cyclones.

There are good records of cyclones because they do much damage and so are recorded. Prior to satellite recording, a strong tropical depression over a heavily populated, low-lying area might cause much damage and be widely reported, whereas a much more severe cyclone that remains over the ocean may have not been noticed at all. Consequently, it becomes very difficult to assess long-term trends in cyclone frequency and intensity. The records of the number of tropical cyclones each year in the Atlantic (Figure 5.9) could be construed to show an increase over the past 150 years, and the changing incidence since 1900 has been related to sea-surface temperature.[10] A wider-ranging study concluded that there was no trend in the number of tropical storms over the period 1970–2004, but that there had been an increase in the number of intense cyclones.[11]

Records held by the Australian Bureau of Meteorology suggest some decline in the number of tropical cyclones in the Australian region since 1970 but an increase in the number of severe cyclones. Over the tropical

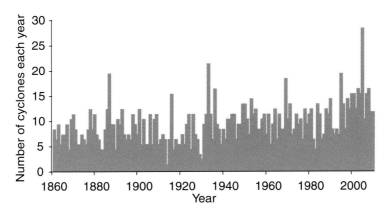

Figure 5.9 Atlantic cyclone count from 1860.[12]

north-western Pacific, typhoon records since 1960 show variation but no particular trend.[13] There is a 1000-year record of typhoons striking Quandong Province in China, but over all this time there does not appear to be any particular change in the frequency of these storms.[14]

A study by Jon Nott and colleagues at James Cook University in Cairns demonstrates how careful scientific investigation is able to decipher unexpected historical detail.[15] Cyclones dump a lot of rain and, as they progress, the isotopic composition of their rainwater changes subtly (box: Isotopes). After they make landfall and then track across north Queensland, their rain grows more and more distinctive. Some of the rain seeps into the ground, and in particular at Chillagoe in north Queensland, down into limestone caves. Each wet season the cave water drips from stalactites in the cave, leaving behind a thin coating of limestone. By analysing successive seasonal layers of limestone on a stalactite, these scientists were able to identify years when the ground water carried the distinctive signature of a cyclone. There was no indication of any change in the frequency of cyclones over the past 100 years compared to the preceding 800 years.

ISOTOPES

Atoms were once thought to be the indivisible particles of all matter. That idea fell apart when it was discovered that atoms were made up of a positively charged nucleus having more than 99.8 per cent of the atom's mass, surrounded by negatively charged electrons. Then the nucleus itself was found to contain two types of particle: protons, which carry the charge, and equally massive, charge-neutral particles called neutrons. The number of protons (and the equal number of electrons) determine the atom's chemical behaviour and defines the element, and the protons plus the neutrons determine its weight. The sum of the protons and neutrons is called the 'atomic weight'.

Hydrogen is the simplest atom: it has one proton and one electron and on the scale of atoms, its weight is 1. There is a rare form of hydrogen having a neutron in addition to the proton in its nucleus, called deuterium. It behaves in the same way as hydrogen chemically, but has twice the weight. An even rarer variety has two neutrons and is called tritium, used in making hydrogen bombs. These three varieties of hydrogen are called isotopes. All other types of atom have isotopes, but the three isotopes of hydrogen are the only ones that have individual names.

Carbon has three isotopes. Of all the world's carbon, 99 per cent has 6 protons and 6 neutrons, and so an atomic weight of 12. The other 1 per cent has an extra neutron. Rather than giving these two separate names, they are denoted ^{12}C and ^{13}C respectively. There is a third isotope of carbon, ^{14}C, created in the upper atmosphere out of nitrogen by the impact of cosmic radiation. It is a radioactive isotope, and used for radiocarbon dating.

Oxygen also has three naturally occurring isotopes. Most (99.7 per cent) has 8 protons and electrons and 8 neutrons, ^{16}O. Much less abundant at 0.02 per cent, but important to climate science is ^{18}O, having 10 neutrons. As ocean water is warmed, the heavier $H_2^{18}O$ water molecules do not evaporate quite as easily as the light $H_2^{16}O$ ones, so the water vapour is a bit richer in $H_2^{16}O$ than the ocean, and the difference depends on the temperature. Then, when it rains, the opposite happens. The heavier $H_2^{18}O$ condenses from water vapour to water drops proportionately first, which means that as a rain depression sweeps across, the early rain contains more $H_2^{18}O$ than later storms. All put together, by measuring the amounts of heavy and light oxygen isotopes in a fossil shell, scientists can work out the temperature of the water where the shell lived. By measuring the isotopic composition of ice the temperature of the snowfall can be found.

DROUGHT

Drought and cyclone: opposite in terms of rainfall, similar in being difficult to analyse statistically because they are infrequent. Drought is defined as 'insufficient water to meet needs'. This definition is clearly a human construct, for as we saw from George Goyder's work, the native plants that grow abundantly in a region a farmer describes as drought-prone have quite sufficient water for their needs. One could argue that Australia does not suffer from drought; it suffers from the attempt to grow inappropriate plants.

Charles Darwin must have recognised this. During his 18-day stop in Sydney in January 1836, he wrote in his diary:

> Nowhere is an appearance of verdure & fertility, but rather that of arid sterility: – I cannot imagine a more complete contrast in every respect than the forest of Valdivia or Chiloe, with the woods of Australia … Although this is such a flourishing country, the appearance of infertility is to a certain extent the truth; the soil without doubt is good, but there is so great a deficiency in rain & running water, that it cannot produce much. The pasture everywhere is so thin, that already Settlers have pushed far into the interior; moreover very far inland the country appears to become less profitable. – I have before said, Agriculture can never succeed on a very extended scale.[16]

Nobody in England listened; the political need for the colony to succeed would not allow for any news but good news to reach home. In the 1920s the geographer Griffith Taylor tried to give the same message. By then it was generally thought that Australia could support a population of 500 million inhabitants. Taylor argued that this was a drought-prone country, and that 20 million was the limit.[17] His 1911 book, *Australia in its Physiographic and Economic Aspects*, was initially banned in Western Australia, and eventually the sceptics forced him out of his university job and out of Australia.

But accepting the farmer's definition, Australian drought records do not seem to show much long-term change, though the first decade of this century does look to be rather more under the influence of drought than any earlier decade. Droughts are compared across regions and time, using the Palmer Drought Severity Index, a standardised measure ranging from about minus 10 (dry) to 10 (wet), of surface moisture conditions. Figure 5.10 (p. 84) shows this index for the Riverina district in south-eastern Australia, while Australia's major droughts are shown also as yellow markers.

Caroline Ummenhofer and colleagues at the University of New South Wales analysed the occurrence of droughts in south-eastern Australia, and linked them to changes in the Indian Ocean.[18] They concluded:

> When taken in the context of other historic droughts over the past 120 years, the 'Big Dry' (the drought from 1998) is still exceptional in its severity … Furthermore, the severity of the 'Big Dry' has been exacerbated by recent warmer air temperatures over the past few decades. Warmer air temperatures lead to

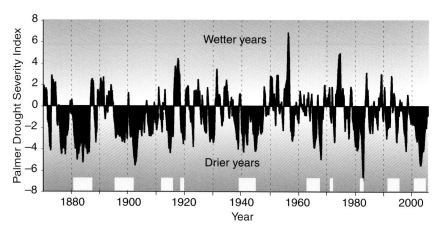

Figure 5.10 Drought years in the Riverina district of south-eastern Australia up to 2005. Drier years are marked by the black columns below the line, and compared with national droughts shown at the bottom in yellow. Higher negative values indicate more severe droughts.

increased evaporation, which further reduces soil moisture and worsens the drought. While this work does not explicitly focus on the link between changes (in the Indian ocean) and recent regional and global warming, it does send a stark message: in a warmer world, the severity of droughts would likely become far worse.

The story in other countries is similar in not revealing any systematic change in drought frequency. Piechota and colleagues used historical records and analysis of tree rings to establish frequency of drought for the Upper Colorado River in the United States.[19] Prior to 1900 the pattern appears to be random, then the results suggest the 20th century has been much less drought-prone than earlier centuries, though at the time the paper was written (2004) the region was experiencing the most severe drought in 80 years.

Studies undertaken in Europe, Africa and China, while generally agreeing with the rainfall changes of the past 100 years shown earlier, find little evidence of any long-term changes in drought severity or frequency.

In 2004, three scientists from the National Center for Atmospheric Research in Boulder, Colorado,[20] published their analysis of worldwide soil moisture changes from 1870 (see Figure 5.11). They showed that since 1950 there has been a tendency for more severe droughts, most notably across the Sahel of Africa, and for heavier rainfalls in parts of North and South America. As is hardly surprising, these results mirror

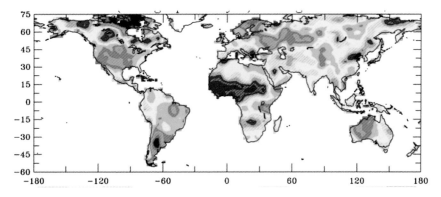

Figure 5.11 Global trends in the Palmer Drought Severity Index since 1950.[19] In this figure, reds and oranges are areas that have tended to experience more severe drought, while darker greens and blues are regions that have become wetter. Source: © American Meteorological Society.

the worldwide rainfall compilations and emphasise that climate change of some kind is a reality.

In early 2010, the *Canberra Times* newspaper reported that there was no evidence to link drought to climate change in eastern Australia. This was a statement attributed to David Post, one of the authors of a CSIRO–Bureau of Meteorology report that had been published in September of the year before.[21] There was immediate comment. Australian Greens Senator Bob Brown was quoted as having accused the CSIRO of 'caving in to political pressure' to soften its stance on climate change in the lead-up to that year's federal election. 'We should ask why CSIRO is prepared to turn an unaccountable blind eye to recent climate trends in Tasmania. This undercurrent of scepticism would seem to suggest the report has been politicised,' Senator Brown said.[22]

But that is not what the CSIRO and the Bureau of Meteorology did. In preparing for the research that underpinned the report, the scientists explored three scenarios:

1) To model river flows using all of Tasmania's previous 84 years of record.

2) To model river flows under the drier conditions of the previous 11 years, to see what might be predictable if this WAS an ongoing response to global climate change.

3) To model projected river flows to 2030, using past records and global climate change predictions. Three assumptions were compared with the recent past, one with the climate being wetter, a second with it

being much the same and a third with a drier climate to see what to expect if these drought years were just abnormal or the shape of things to come.

Weather is not climate. If a long drought is a weather event rather than a response to climate change, it would be a mistake to regard it as something special. So the report did not presume there was a link and David Post's comments were typical of a cautious, knowledgeable scientist.

FLOODS

Unlike droughts, about which we can do nothing, floods can be mitigated by engineering. That means it is not easy to compare floods over recorded history, since the rain that causes a severe flood in one century may do no more than fill the new control dams in the next. The places to look for evidence of any change in flood frequency or magnitude would not be in the heavily populated parts of the world like Europe, but perhaps in the more sparsely populated Australia.

The other side of that coin is that Europe has excellent and long records, so at least there are data to be considered. Such long records are held for both the Atlantic and Mediterranean basins of Spain. In those regions, the early 17th century and the late 18th century, continuing through to the 19th century had significantly more floods than other eras, with a general decline through the 20th century.[23] The River Vistula in Poland experienced more floods from 1400 to 1600 than in the following three centuries, with an increase evident in the 20th century.[24] German tributaries to the Danube show essentially random flood patterns since 1350.[25]

A 19-member team of scientists compiled records of European floods back to 1500.[26] Writing in the journal *Climatic Change*, they concluded: 'Most significantly, recent changes in the variability of flood frequencies are not exceptional if compared to the flood frequency of the past 500 years and show no overall trend similar to the widely-cited "hockey-stick" trend for temperatures.'

Australia cannot boast such long records, but there have been spectacular floods since European settlement, many across southern Queensland and northern New South Wales. Such floods commonly occur when tropical lows cross the Queensland coast. If they rain out before crossing the Great Divide, which is typically only a 100 or so kilometres inland, most of the water runs to the sea, flooding the immediate neighbourhood.

Not far inland from Brisbane is the city of Ipswich, on the Bremer River. Flooding of the Bremer has been recorded since 1850 and the records show an increase in flood frequency toward the latter half of the 20th century.[27] But the Brisbane River, into which the Bremer flows, shows the reverse trend. Perhaps that is because there was a big flood in 1974, and since then control of the Brisbane River has improved so that floods did not affect the lower reaches and the city of Brisbane so badly.

All this changed in January 2011. The second-strongest *La Niña* ever, coupled with unusual strength in the annual monsoon, gave Queensland a record wet December, and indeed for the whole of Australia the period July to December 2010 was the wettest on record. In the weeks leading up to Christmas, flooding was widespread across Queensland, continuing into the first week of January. With the ground already thoroughly saturated, and rivers and dams full, three days of unprecedented rain across the Brisbane hinterland dumped more than 600 millimetres in some areas, with local falls reaching 60 millimetres in an hour. This water poured in what was described as 'an inland tsunami' down the Lockyer River, destroying the township of Murphy's Creek as well as flooding the Bremer River and central Ipswich. The water continued into the Brisbane River until all the low-lying parts of Brisbane itself were flooded.

The story is different if a storm drops a lot of rain west of the divide. The land is almost flat from Queensland to South Australia. Heavy rain from tropical storms takes three months to reach its ultimate destination. Very quickly, the rivers overflow their banks and the water spreads for many kilometres across the countryside, such as is shown in Figure 5.12 (p. 88). Across south-central Queensland and north-central New South Wales, the water flows into the Darling River and thence to the Murray. Further west, big floods eventually reach Lake Eyre, normally a salt-encrusted depression 240 metres below sea level.

In any particular river catchment, floods are quite infrequent, but by combining flood records for a localised group of major rivers across western Queensland and New South Wales, some indication of frequency can be estimated.[28] Adding flow records for Cooper Creek, the Warrego and Paroo rivers to records of the filling of Lake Eyre provides a check on the individual river flood events. It shows (Figure 5.13, p. 89) that floods happen intermittently, with perhaps an increased frequency in the past 25 years; sometimes large and widespread, sometimes more local and sometimes, as in February to March 2010 and again in February 2012, spectacular to look at, catastrophic to experience.

Figure 5.12 The Darling River at Bourke, NSW, during the flood of 1971. Source: John Brooke.

Throughout this chapter the emphasis has been on rain (or lack of it) in Australia. Being an island continent, Australia is uniquely placed to be sensitive to the weather or climate effects of the ocean. It seems to be clear that extreme changes in rainfall, such as droughts, floods and cyclones, are at least connected to complex oscillations in a dynamic ocean–atmospheric system. Of these, the *El Niño* Southern Oscillation and the Indian Ocean Dipole seem to have a significant influence on Australia. The floods of just those three Australian rivers and the filling of Lake Eyre follow the *La Niña* phase of the SOI with almost metronomic precision.

There is some evidence that *El Niño* events have become more frequent over the past 30 years. According to analyses by the Australian Bureau of Meteorology and the University of East Anglia's Climatic Research Centre in England, in the 30-year period since 1980 there have been five *El Niño* events; in the 90 years before 1980 there were also only five. To a certain extent this comparison depends on the definition of an *El Niño* event. Selecting years when the average SOI value was less than minus 10 conforms to the above analysis (see Figure 5.13). The very prolonged negative phase of the SOI from 1990 to 1995 adds to the suggestion that *El Niño* has become more frequent since 1980.

A SHORT INTRODUCTION TO CLIMATE CHANGE

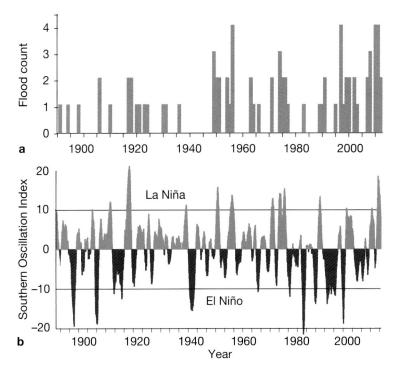

Figure 5.13 a) Years when there were floods on Cooper Creek and/or the Warrego and the Paroo rivers and/or Lake Eyre. b) Compared to the 13-month averaged SOI. *El Niño* events are negative, *La Niña* events positive.

Although *El Niño* events are generally associated with reduced rainfall or drought in eastern Australia and Indonesia, and *La Niña* with more rain,[29] the Bureau of Meteorology observes that as an *El Niño* event decays, good rainfalls sometimes occur. This happened in 1973, 1983, 1995, 1998, 2003, 2007 and spectacularly with unprecedented flooding of Queensland in March 2010, again as an *El Niño* was fading.

From the results described in this chapter it is evident that rainfall patterns have changed over the century. For Australia, the changes are quite marked, with an average 5 millimetres per year decline in eastern Australia since 1971, and an increase of similar magnitude over northern Western Australia. Globally, there are also changes, similar in extent to those of Australia, and similar in there having been both increases and decreases in different parts of the world.

Cyclones, as far as I can see, do not reflect a single trend and possibly no trend at all. The evidence in the literature does not seem to me to support the hypothesis that global warming might have caused more destructive and more frequent cyclones. Compared to daily measurements of temperature and rainfall, cyclone 'measurements' (events) are highly infrequent. Possibly, the numbers show no trend because there is no trend; possibly they show no trend because there are too few observations to establish anything meaningful. However, droughts and floods in eastern Australia do appear to have become more extreme over the past 50 years.

SUMMARY

Topic	Key statistic	Conclusion
Rain	Global rainfall variations are of the order of ±50 mm/year since 1970.	Changing temperatures have changed rainfall patterns.
Cyclones	Worldwide, no evidence for a change in frequency.	Possible small increase in intensity since 1980.
Drought	Global variation seems to average out.	No evidence of long-term change globally, possibly more frequent in Australia.
Floods	Too infrequent to show a clear signal.	No evidence of long-term change globally, possibly more frequent in Australia.

Of all the fresh water in the world, the water sourced by rain and feeding the rivers, lakes and ground waters makes up only about one-third; all the rest is frozen, most of it in the vast Antarctic ice sheet. In the next chapter we will investigate the snows and glaciers of the high mountains, and the ice sheets of the polar regions.

FURTHER READING

Fagan B (2009) *Floods, Famines and Emperors*. Basic Books.

Nott J (2006) *Extreme Events: A Physical Reconstruction and Risk Assessment*. Cambridge University Press.

SNOW
AND ICE

Like an army defeated
The snow hath retreated ...

William Wordsworth

In the mid-18th century, the grand tour of Europe was the aim of many wealthy young Englishmen, and no tour was complete without a visit to at least one of the famous Swiss mountain glaciers. The glaciers had been made a tourist icon by an amazing 1.85 metre-wide photograph of the Lower Aar Glacier taken in 1858 by Auguste-Rosalie Bisson and shown at the Paris Exposition of that year. The photograph showed the fractured and crevassed glacier winding its way down the valley.

Stereo-pair photographs were also popular in the 1850s, and the glaciers did not miss out. The Lower Grindelwald Glacier was photographed in stereo in 1858, by which time it was known that glaciers actually moved. The pioneering naturalist Louis Agassiz had discovered this when he saw how a hut built on a pile of rocks in the middle of a glacier had moved downhill. Agassiz also understood that great piles of rocks left across mountain valleys, often acting as natural dams, marked where the end, or snout, of a glacier had been in earlier times.

Figure 6.1 a) The Lower Grindelwald Glacier in 1858. Source: The collection of Richard Wolf, Fribourg. b) Recent (2003) view of the same part of the glacier gorge. Source: Daniel Steiner.

Agassiz's recognition of glacial deposits in mountains eventually led him to conjecture that similar rock deposits across Europe where there were no mountains must also have had a glacial origin. He boldly concluded that great masses of ice had once covered the land. But nowhere in *The Holy Bible* was there any mention of an 'ice age'; Agassiz's theory was met with denial by those whose beliefs or imagination would not allow them to conceive of such a thing.

South of the Arctic Circle, mountain glaciers are not very many degrees below freezing and so they are highly susceptible to small temperature changes. A mountain glacier's lower end, its snout, is at a place where it melts at the same rate as it is being pushed downhill by the ice behind it. If summer temperatures fall for a few years, melting decreases and the snout moves downhill some distance until melting again balances the flow of the ice. If warmer summers continue for a few years, the snout melts away until it is far enough uphill, where average temperatures are cooler, to again reach a balance. Such uphill movement of the snout is termed 'retreat', even though the mass of the glacier continues to move downhill. Increased temperatures will certainly accelerate melting, but decreased snowfall will reduce the rate of the glacier's advance just as increased snowfall will increase it. If reduced snowfall is the result of drier air over a long period, this will also contribute to loss of ice by direct evaporation. Particularly in dry air, ice can evaporate – termed 'sublimation' – without necessarily melting. Snout retreat then does not necessarily indicate warming.

The Lower Grindelwald Glacier has been recorded in drawings and photographs since 1535, which allows a good record of the position of the snout from that time. According to the Swiss scientist Steiner and his colleagues, the glacier reached its maximum known length about the year 1650. It retreated by up to 500 metres during the 1700s, then returned to its earlier maximum length in 1850, shown in the photograph at Figure 6.1a. Over the following 150 years it steadily retreated. Today, if you now want to photograph the snout you have to walk a further two kilometres up the valley and climb 325 metres higher.[1]

As anyone who has walked in the Swiss Alps or any other mountains knows, it gets colder the higher you climb. The rate at which the temperature drops, called the 'lapse rate', is 6.5°C per 1000 metres. The movement uphill of the snout of the Lower Grindelwald Glacier might seem to suggest that the temperature there has risen by about 2°C in 150 years, but equally possibly there might just have been less snow. Either way, the Lower Grindelwald Glacier seems to be telling us that the Swiss climate has changed in 150 years.

A hundred years after Agassiz and the grand tours of Europe, Ernest Hemingway and his wife visited Africa. His short story (1936) and later the movies *The Snows of Kilimanjaro* (1952 and 2011) made the glaciers and ice of this volcanic peak familiar sights. If you want to see the snows, go soon. Since they were first mapped in 1912, the area of snow and ice has contracted, and 85 per cent had gone by 2007, leaving an area of only 1.85 square kilometres (Figure 6.2b). As well as shrinking by melting, the ice is rapidly thinning as a result of sublimation. Professor Thompson and his colleagues from the Byrd Polar Research Center of Ohio State University in the United States predicted in 2003[2] that if the loss of ice continues at the present rate, all would be gone by about 2030.

There is disagreement about the cause of the shrinking Kilimanjaro glaciers. Thompson and his colleagues considered that drier air, changed

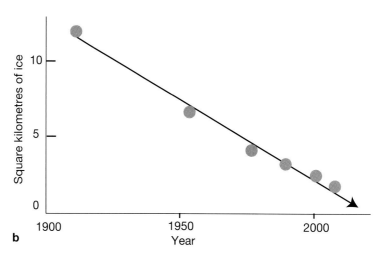

Figure 6.2 a) Mt Kilimanjaro in 1996. Source: Roger Hambly. b) The decreasing snow cover of Mt Kilimanjaro. Source: Thompson et al., 2009.[3]

land use and warming have all played a part. From the Tropical Glacier Group of the University of Innsbruck in Austria, Thomas Mölg and his colleagues published their conclusions after detailed modelling of these glaciers.[4] They believed that reduction in humidity and snowfall were largely responsible, with drier air leading to greater sublimation of the ice and reduced snowfall failing to restore the ice lost to the Sun's heat. Mölg and colleagues did not believe temperature increase over the past 30 years to be an important factor.

According to Māori legend in New Zealand, Hinehukatere was a young mountain woman. She spent her days wandering the high alps, sometimes accompanied by her lover, Tawe. Unfortunately, Tawe was not so knowledgeable about high mountains, and he was caught in an avalanche. Inconsolable, Hinehukatere cried and cried until her tears became a stream, which froze as it flowed down the mountain. Thus, New Zealand's largest and best-known glacier, Franz Josef, remains as the evidence of this Māori legend.

For those who love mountain glaciers, the analogy of frozen tears is apt. The Franz Josef Glacier, as Europeans named *Ka Roimata o Hinehukatere* ('the tears of Hinehukatere') was in retreat until 1982, but from then until 2002 at least it advanced, so perhaps there is no need to cry over this glacier and the adjacent Fox Glacier. Nonetheless, as Endre Før Gjermundsen's master's thesis at the University of Oslo showed,[5] overall New Zealand's glaciers retreated by about 17 per cent between 1978 and 2002. A more extensive study of New Zealand glaciers found a 50 per cent loss of glacier area from 1850 to 1970, after which, like the Franz Josef, they advanced, particularly during the 1990s. Since then, there has been an even more rapid retreat, and according to the New Zealand National Institute of Water and Atmospheric Research:

> This was mainly due to the combination of above normal temperatures and near normal or below normal rainfall for the Southern Alps during winter, and La Nina-like patterns producing more northerly flows creating normal-to-above normal temperatures, above normal sunshine, and well below normal precipitation for the Southern Alps particularly during late summer.[5]

According to Hoezle and colleagues,[6] mountain glaciers in Europe and New Zealand both experienced mass losses of the order of 50 per cent between 1870 and 1970. From 1970 to 2000, the New Zealand glaciers advanced again, whereas those of Europe continued to retreat. The same story can be found in almost all mountain glaciers: in the Andes in South America, one glacier is now retreating more than 30 times faster than it did in 1960. In the United States, the number of glaciers in Glacier National Park decreased from 150 to 37 over the past 150 years.[7]

Of all the world's mountain ranges, the Himalayas are richest in glaciers. Several studies of the extent of the Himalayan glaciers have been made, including one published in 2007 by a group of Indian scientists. They compared the area of 466 glaciers as measured in 1962 by aerial photography and in 2007 by satellite imagery. There has been a general decline of about 20 per cent over that period.[8] A detailed analysis of the impact of climate change on Himalayan glaciers was published in the same year by the International Centre for Integrated Mountain Development. No glaciers were reported to be advancing; glacier retreat rates ranged from 10 and 60 metres a year.[9]

However, a short report in 2008 about a group of 250 glaciers in Pakistan's Karakoram region of the western Himalaya region found that 65 per cent were either advancing or stationary.[10] The authors of this work attribute this to increased snowfall in that region. You may have read of a statement in the IPCC's Fourth Assessment report that Himalayan glaciers were vanishing at an alarming rate and would be gone by the year 2035. This was an error, and it has been explained and corrected. But as a news feature in the journal *Nature Climate Change* (March 2010) put it:

> One thing is clear: the glaciers won't vanish by 2035, as the Intergovernmental Panel on Climate Change (IPCC) claimed in its 2007 assessment report. This error and others in the IPCC report's section on Himalayan glaciers – widely reported elsewhere – have now been corrected. But the ensuing furore has highlighted how little is actually known about the fate of glaciers in this region. The errors 'were mainly based on the desire to say something', says glaciologist Richard Armstrong of the National Snow and Ice Data Center in Boulder, Colorado. 'But you need to know that if there's no data, you shouldn't say anything.'

An even broader study of mountain glaciers by Mark Dyurgerov and Mark Meier at the Institute of Arctic and Alpine Research at the University of Colorado published in 2005 considered the mass changes of glaciers in Alaska, the Arctic, Europe, the Himalayas, the United States, Canada, the Andes and in Patagonia.[11] Their results are shown in Figure 6.3. Dyurgerov and Meier found a slight increase in European glacier

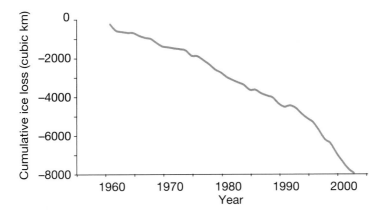

Figure 6.3 Cumulative loss of ice over the period 1960 to 2005 from all the world's glaciers, excluding the icecaps of Antarctica and Greenland. Ice loss is measured in cubic kilometres of ice. Source: Dyurgerov & Meier, 2005.

mass until 1995, but a loss of mass in all other regions, accelerating since about 1995.

These authors also estimated the rise in sea level resulting from glaciers melting, and that result is shown in Figure 6.4. Super-imposed upon the sea-level graph is the HadCrut3 global temperature curve. The match is quite obvious, though in all such comparisons one should not automatically assume cause and effect.

Changes detected at any particular place may well be the result of specific local conditions, and taken alone may not be globally significant.

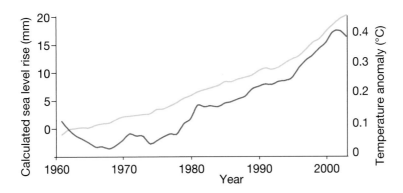

Figure 6.4 Calculated rise in sea level as a consequence of melting glaciers (pale line) compared to global temperature change (dark line). Source: data for sea-level rise from Dyurgarov & Meier, 2005.

It is only when many lines of evidence point in the same direction that one is forced to sit up and take notice. Overall – and that includes the fringing glaciers of Greenland and Antarctica but not their huge ice-caps – there has been a decline in glacier volume between 1960 and 2004, amounting to 8000 cubic kilometres of ice, or 3 per cent of the total volume of valley glaciers.

The advance and retreat of mountain glaciers is nothing new. In Alaska, glaciers have advanced and retreated four times in the past 1000 years, at least partly in response to changes in the Sun's intensity.[12] Ten of Norway's glaciers have been documented by Nesje and colleagues based on geological observations of deposits left for the past 11 000 years.[13] All the glaciers disappeared about 8000 years ago in response to summer temperatures thought to be 1.5–2.0°C warmer than today's. At about 5000 years ago the glaciers returned, and variously advanced and retreated, with a strong retreat about 1000 years ago followed by their greatest advance during the Little Ice Age, 200 to 300 years ago. All these glaciers are now in retreat, though some show a net mass gain as a result of increased snowfall more than balancing the loss by melting at the snout. The overall situation in 2002 was summarised by Roger Braithwaite and Sarah Raper from the University of Manchester in England and the Alfred Wegener Institute for Polar Research in Bremerhaven, Germany, respectively, when they wrote: 'Increased melting of glaciers and ice caps, excluding Greenland and Antarctica, will probably represent the second largest contribution to global sea level rise by 2100.'[14]

A large international team of researchers led by Paul Mayewski from the University of Maine in the United States has compiled hundreds of individual records revealing several episodes of rapid climate change since the end of the last ice age, about 12 000 years ago.[15] In this context, 'rapid' refers to cooling periods lasting between 400 and 1000 years. Mayewski and colleagues found six times when glaciers advanced, with simultaneous aridity in equatorial regions. The only temperature records for this period come from Antarctic and Greenland ice-cores, and they indicate a total range of perhaps 1°C either side of the present temperature.

It is important to put these changes and our present rate of global warming into perspective. The 'rapid' changes of the past 12 000 years amount to a cooling or warming of 1 to 2 °C over as much as 1000 years. These changes were sufficient to cause alpine glaciers to advance, retreat or disappear. They provide a very clear indication of what will happen to our ice if the temperature continues to rise at its 20th-century rate of 7°C per 1000 years. This is not a 'rapid' climate change; it is an avalanche.

PERMAFROST

North of the Arctic Circle, across Siberia, Alaska and northern Canada, the ground is frozen solid throughout the year, to a depth of up to 1 kilometre. During the Arctic summer, across the tundra there is usually enough surface melting for mosses, lichens and dwarf shrubs to grow. This vast region of frozen ground, known as 'permafrost', has existed for many thousands of years. But now the Arctic permafrost is warming.[16] Measurements made at 10 locations around the Arctic – two in Russia, two in Canada, five in Alaska and one in Norway – all show an overall gradual rise in annual average temperature of the permafrost. Really cold permafrost, such as that measured at Alert in Canada, is warming at 1°C per decade. Permafrost at a temperature just below zero, warms slowly because the Sun's energy is used up in melting, rather than raising the temperature. The temperature of permafrost cannot rise above zero until all the ice has melted (and it is no longer permafrost).

The rate of warming of the permafrost is alarming, but is exactly in agreement with climate scientists' expectation for the way the temperature should rise with increasing amounts of greenhouse gases. Based on the pattern of 20th-century global warming and evidence from the geologic past[17] (see Chapter 8), the Arctic regions are predicted to warm faster than the global average, and these observations seem to agree with that prediction. An indirect indication of Arctic climate change is evident in an analysis of the flow of six major rivers draining the Eurasian Arctic land mass into the Arctic Ocean. Since 1960 the total flow from these rivers has been increasing at a rate of about 2 cubic kilometres of water each year.[18]

ICECAPS AND POLAR SEA-ICE

Reduction in the area of the Arctic sea-ice received much media attention during the first decade of the 21st century, with the plight of polar bears very much in the forefront of news and documentaries. The area of Arctic sea-ice can be estimated from historical records over the past 150 years and, more recently, measured from satellite photographs. Both the areas of maximum and minimum sea-ice have declined over that period. From the satellite data alone, by the end of the summer of 2011 the Arctic sea-ice cover had declined by 40 per cent from its 1979–89 average, as Figure 6.5 (p. 100) shows.

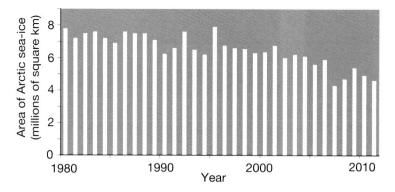

Figure 6.5 Changes in Arctic sea-ice area since 1979. Source: US National Snow and Ice Data Center.

Submarine records from 1980 and recent satellite data combine to show that over the past 30 years the thickness of Arctic sea-ice has halved.[19] From an average winter thickness in 1980 of about 3.7 metres, the ice had thinned by 2008 to 1.7 metres, with 0.5 metres of that loss being since 2004.

By analysing Arctic ocean sediment, a joint Canadian and US research team have concluded that over the past 9000 years the area of Arctic sea-ice has fluctuated considerably, in some eras being lower than it is today, with sea-surface temperatures being several degrees warmer.[20]

Figure 6.6 Antarctic glacier cracking as it reaches the sea. Source: Giselle Coromandel.

A SHORT INTRODUCTION TO CLIMATE CHANGE

In contrast to the Arctic sea-ice, the Antarctic sea-ice coverage has increased by 1 per cent a decade over the past 30 years,[21] whereas Antarctic continental ice has been lost. An international team of scientists writing in the journal *Nature* in 2007 concluded that the ice sheet as a whole was certainly losing mass.[22]

Since then, several teams of scientists have analysed the Antarctic ice sheet using different methods: radar estimates of ice-volume change, or gravity measurements using the Gravity Recovery And Climate Experiment satellite (GRACE) to measure mass loss. All found an overall mass loss, though the annual amounts estimated ranged from 30[23] to 250[24] cubic kilometres. Another team from the University of Texas in the United States studied the ice-mass balance at several locations in Antarctica and found significant mass losses from the Antarctic Peninsula and the western coast of Antarctica, and a lesser but significant increase in total ice at Enderby Land and in two other places.[25]

Another analysis of ice loss in just the Amundsen Sea catchment estimated about 40 cubic kilometres of ice was lost in 2009.[26] From the perspective of 'that's not much', using the larger figure (250 cubic kilometres a year) it would take 120 000 years before all the Antarctic ice melted. From the 'that's a lot' perspective, enough ice melted in 2009 to make an ice block one kilometre wide and one kilometre high, stretching the distance between Canberra and Sydney.

An example of the decline in Antarctic ice can be seen using an open-access website maintained by the Research School of Earth Sciences at the Australian National University in Canberra (see http://grace.anu.edu.au). By selecting a point on the globe, over a given time period since 2003 the change in ice, expressed as millimetres of water, can be visualised. This is not a computer model; it is processed real data obtained from the GRACE satellite (Figure 6.7, p. 102).

Differences in interpretation of results highlight the very essence of how science works. Particularly with the development of new instrumentation, early results may well show the direction, then as the methods improve, details emerge that were not visible at first. In the 17th century, Italian astronomer Galileo's first observations of Saturn led him to describe it as having two large moons, often referred to as 'ears'. Later, the use of better telescopes revealed the rings of Saturn, and since then space craft have explored their make-up. Galileo was not 'wrong'; he interpreted the remarkable images he saw in his small telescope as best he could, and later work refined his interpretation.

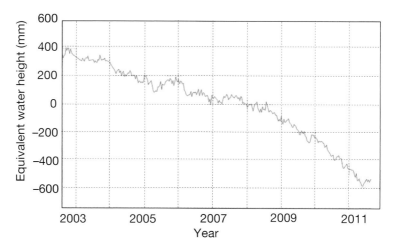

Figure 6.7 Ice loss at a point in Antarctica (latititude −76.51, longitude −129.07), measured from the GRACE satellite and processed at the Australian National University.

The Greenland icecap is the world's second-largest after Antarctica, and its behaviour has been the subject of many studies. One such from the University of Texas[27] estimates an annual melting of about 240 cubic kilometres of ice over the period 2002–5. Another study published in 2008 found that there has been a marked loss of ice since 1990, such that by 2007 the ice sheet was vanishing almost three times as fast it was in the 1960s.[28] As with the Antarctic icecap, by 2009 the GRACE satellite had also found accelerating ice loss from Greenland, such that in the last two years of the study it shed a bit less than 300 cubic kilometres of ice per year.[29] This alone is enough to raise the global sea level by 1 millimetre a year.

Using radar altimetry, a team of scientists from Norway, Russia and the United States[30] looked at changes in the elevation of the Greenland ice sheet over an 11-year period ending in 2003. These authors pointed out that they were unable to estimate ice-volume changes and further that the 11-year period of the study was too short to establish long-term trends. Nevertheless, they found that in the high interior of Greenland the elevation had increased by about 60 centimetres, implying increased snow accumulation. Below 1500 metres, melting and thinning of the ice has resulted in a loss of about 20 centimetres. In short, more snow is falling higher up and more ice is melting lower down.

These apparently opposite results may be confusing. As we saw in Chapter 5, as the ocean temperature rises, more water evaporates and the atmosphere acquires more water vapour. This will be rained out somewhere, and it turns out that Greenland is one such place. In mid-summer, the maximum temperature ever reached in the highlands is minus 10°C. Extra precipitation there would fall as snow even if the world's average temperature rose by 5°C; the 3000 metre-high Greenland icecap would still be below freezing. Around the coastline, the story is different. There, the summer maximum may reach 12°C. Therefore, the highlands acquire more snow because of global warming, but the lowland glaciers are melting more for the same reason.

In 2011, the total amount of ice lost to Antarctica and Greenland was measured by the same scientists who made the earlier analyses.[31] The rate at which ice was melting from these two icecaps was then accelerating by about 36 cubic kilometres a year annually, so that the 2011 loss amounted to some 600 cubic kilometres. If that acceleration is maintained, sea level will rise 15 centimetres by 2050 from this alone.

The Devon Island Ice Cap in Canada is one of the largest after Antarctica and Greenland, containing almost 4000 cubic kilometres of ice. Between 1960 and 1999, this icecap lost more than 70 cubic kilometres of ice and the rate of melting has increased since 1985.[32] This is yet another place where ice is melting faster than it is being replaced.[33]

From all of these studies it is clear that each year, more ice is lost than in the previous year. There is a lot of ice in Antarctica, about 30 million cubic kilometres, yet if its *accelerating* loss were to continue unchanged, it would take only 2000 years or so to lose all of Antarctica's ice. The last time Antarctica was ice free was about 40 million years ago. Change is certainly happening rather more quickly than it ever did in the geological past.

CONCLUSIONS

Mountain glaciers are said to be to climate change as the canary in the cage is to coal gas. Because a glacier's snout is positioned at a balance point between melting and advancing of ice down a valley, small and consistent changes in temperature, particularly summer temperature, soon produce a shift in the position of the snout. Almost all recorded mountain glaciers, as well as those shedding from northern hemisphere icecaps, have been receding or losing mass for most of the 20th century and, taken in total, the loss of ice closely follows the global temperature rise. The decline in the area of Arctic sea ice each summer is a second phenomenon that looks to be related to rising temperatures. The Antarctic, Greenland and

Canadian icecaps are also losing mass. Altogether, the global melting is strong evidence for climate change.

SUMMARY

Topic	Observation	Key statistic	Conclusion
Mountain glaciers	Almost all are receding.	A worldwide decline in glacier volume of 3% from 1960 to 2004.	Temperature rise is melting the glaciers, raising sea level.
Arctic permafrost	Warming everywhere.	Warming as fast as 1°C per year.	The Arctic is warming faster than the rest of the world.
Polar sea-ice	Shrinking quickly in the Arctic, expanding slightly in the Antarctic	Arctic sea-ice cover has fallen 40% since 1980; Antarctic sea ice has increased by 3%.	Rising Arctic temperatures are melting sea-ice.
Icecaps	Thinning in Greenland, the Antarctic and Canada.	250 cubic kilometres lost annually from Antarctica.	Temperature rise is melting the icecaps, raising sea levels.

The last two chapters have been about the Earth's freshwaters: flowing, falling or frozen. Freshwater constitutes only 3 per cent of global water, and most of that is in Antarctica as ice; the other 97 per cent is in the oceans. The atmosphere–ocean system is a vast heat exchanger, and we have already seen one consequence of this in the *El Niño* effect. The ocean is next in our search for evidence for or against climate change.

FURTHER READING

Bajracharya SR, Mool PK & Shrestha BR (2007) *Impact of Climate Change on Himalayan Glaciers and Glacier Lakes.* International Centre for Integrated Mountain Development.

United Nations Environment Program (2007) *Global Outlook for Ice and Snow.*

7

THE
OCEAN

How inappropriate to call this planet Earth when it is quite clearly Ocean.

Arthur C. Clarke

The United Nations Convention on the Law of the Sea proclaims that a country may claim mineral exploration and fishing rights over its continental shelf. In Australia, this is a 200 kilometre-wide zone around the continent, extending to a depth of around 150 metres. Such a claim would hardly have surprised those living 17 000 years ago. In those cold days, the surf broke at the edge of the continental shelf and the hunting rights of early Aboriginal peoples certainly covered all of that 200-kilometre band beyond our present shoreline.

Sea-level rise is one inevitable consequence of global warming, so it is worthwhile knowing what is possible. If the polar ice sheets melt entirely, the Australian shoreline will look like it did in the age of the dinosaurs, 115 million years ago. Then, the ocean covered much of western Queensland and northern South Australia (see Figure 7.1a).

After that, with the slow development of the ice ages, the sea level fell until, in the depths of the last glaciation 25 000 years ago, it would have been possible to walk out to the Great Barrier Reef, or from Melbourne to Hobart, or Queensland to New Guinea (Figure 7.1b). Indeed, in other parts of the world the English Channel and the Bering Strait between Russia and Alaska were dry land. River valleys ran out as far as 200 kilometres beyond the present shoreline. Since then, sea level has risen by well over 100 metres as the world warmed from the last ice age. All the world's great natural harbours, including Sydney Harbour, are simply wide river valleys that were flooded as the ice melted. This is one reason anthropologists have a difficult time establishing the early history of human evolution. People then, as now, liked to live near the sea, where there was good food supply. The rise and fall of the sea level from ice age to ice age kept inundating and destroying any record left by these people, so the majority of the information about those times has been washed away. Though we are now in an interglacial period, with the next ice age estimated to be about 50 000 years in the future, there are still remnants of the great ice sheets. But those remnants are far from trivial. They are the difference between today's landscape and the very different amount of dry land shown in Figure 7.1a.

You can see that sea-level rise and fall is nothing new to the Earth. Most of the last great rise, at a rate of about 10 millimetres a year, was complete 7000 years ago; after that and until 1900 the rise was at the rate of about half a millimetre a year.

Amid all the concern about heat waves and droughts, hurricanes and floods, it is easy to forget that the ocean is second to the Sun in importance in the way the Earth's climate and weather behave. For a start, there

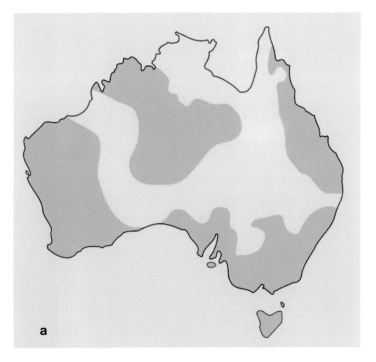

Figure 7.1 a) The Australian continent as it was 115 million years ago, half of it covered by sea.[1] b) The Australian shoreline at the height of the last glacial maximum, 25 000 years ago. Source: after O'Connell et al., 2010.[2]

is an awful lot of ocean. It covers about 70 per cent of the globe, its depth reaches almost 11 kilometres in one place and averages almost 4 kilometres, and it weighs about 300 times as much as the atmosphere.

OCEAN HEAT

The oceans can store a lot of heat. A kilogram of water needs about four times as much heat as a kilogram of the atmosphere does to raise its temperature by 1°C. So roughly, the oceans take in a thousand times as much heat as the atmosphere if they both rise in temperature by 1°C. The oceans do this quite slowly, because the Sun only warms the surface layer and warmer water is lighter (less dense) than the cooler water below and does not readily sink. Waves and ocean currents, especially the Great Ocean Conveyor Belt we looked at in Chapter 3, mix the warm surface water with the cold deep water until eventually the heat drifts down. Oceanographers estimate that it takes around 1000 years for the heat at the surface to be distributed. The temperature is always warmest at the top, but eventually the entire ocean will warm up in response to any rise in global temperature.

HEAT

Heat is a form of energy. When you heat food in a microwave oven you take electrical energy and transform it into microwave radiation. The radiation is absorbed into the food as the radiant energy is turned into heat, causing its temperature to rise. You need that meal, because to keep you heated (alive), you need about 2500 calories a day, or in the units of heat energy, 10 million joules.

At noon, the Sun at the equator puts out 1.36 kilowatts per square metre (see Chapter 3). After reflection of some of that heat, there is perhaps 1 kilowatt left to warm the Earth. Allowing for the reduced impact of the Sun away from the equator and the tropics, the equivalent energy of about one microwave oven for every 3 square metres of surface under the Sun is what drives the Earth's climate!

The ocean warms rather slowly, so not all of it has mixed enough to have responded to any global temperature change of the past 100 years. The estimated change since 1950 in the heat content of the upper 700 metres of the ocean has been analysed by Domingues and colleagues[3] and is shown in Figure 7.2. To put that amount of heat into perspective, the ocean's heat content increased from 1961 to 2000 by about one million times the current annual output of all power stations in the United States (10^{22} joules compared to the United States output of 10^{16} joules).

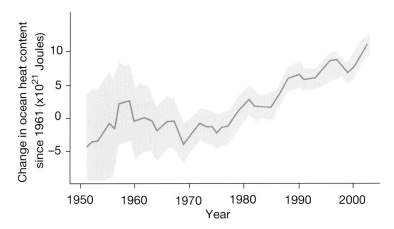

Figure 7.2 Estimates of the change in the heat content of the upper 700 metres of the oceans since 1961, measured in joules x10^{21}. The grey shading indicates the degree of uncertainty of the measurements. Source: after Domingues et al., 2008.[3]

Just like the sea-surface temperature data shown in Chapter 2, there has been a steady and inexorable climb in the ocean's heat content, and therefore its average overall temperature.[4] The actual temperature increase seems almost trivial: 0.1°C in 50 years, compared to 0.5°C for the sea surface over the same period. But that new heat is stored in the oceans and will provide a reservoir of heat for years to come. Even if the external factors warming the surface were to cease right now, the ocean would be there to keep the heat on for the next 100 years. It is already too late to *stop* global warming in our lifetime; we can only limit it.[5]

With a fairly irregular pattern, heat from the oceans is exchanged with the atmosphere. Ocean currents continually mix the warmer surface water with the cooler deep water, but the extent of this mixing varies from year to year. The *El Niño* Southern Oscillation (ENSO) phenomenon is one important contributor; during a *La Niña* episode, when cool water rises in the eastern Pacific Ocean, the global temperature falls, and during *El Niño* years the temperature rises. A very strong *El Nino* event occurred in 1997–8, causing a marked rise in the global temperature. Figure 7.3 shows the global temperature curve above, compared to variations in ENSO below; *El Niño* events are upwards on this graph, *La Niña* are down. When the Pacific Ocean temperature increases a little during an *El Niño* event, as it did in 1998, a few months later so does the global temperature. When the Pacific cools, so does the global temperature, as it did during the 1988 and 2010–11 *La Niña* events. Certainly, the fluctuations in ENSO are mirrored by fluctuations in global temperature, but in addition to the ENSO-driven fluctuations, the temperature shows a steady rise.

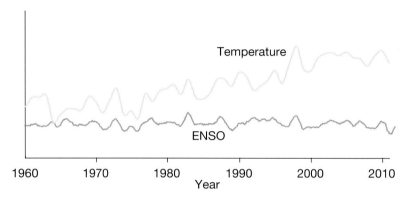

Figure 7.3 Comparison since 1960 between ENSO fluctuations (lower, dark line) and the global land-ocean temperature changes (upper, pale line) (HadCRUT3).[6] The vertical scale is arbitrary and is only intended to show the similarity in timing, not the magnitude of the effect ENSO has on climate.

Understanding this effect is important in the debate about climate change; among other things it has been the subject of a peer-reviewed scientific paper which concluded that global warming is largely driven by ENSO.[7] The authors analysed satellite measurements of temperature in the lower atmosphere (troposphere) over a 27-year period and compared the temperature variation with the Southern Oscillation Index (SOI). They found, as described above, that temperature changes closely follow ENSO changes with a delay of about 7 months, and from this they concluded that as much as 70 per cent of the atmospheric increase in temperature comes from the ocean, not from greenhouse effects.

However, this conclusion turns out to be completely wrong, because McLean and colleagues effectively subtracted the steady annual increase in temperature in the atmosphere.[8] Thus, what they found was the contribution that ENSO makes to annual changes in lower troposphere temperature *after* the annual increase has been subtracted, a well-known amount of about 15–30 per cent of the annual variability.

SEA-LEVEL RISE

An inevitable consequence of warming in the oceans is expansion. Simply by expansion, a 2-metre-deep swimming pool would get almost 2 millimetres deeper if it warmed from 15° to 20°C. As with a swimming pool, so also with the ocean. Like a swimming pool, the ocean is limited by its sides (in this case the continents) so that if the ocean expands its water can only go upwards; that is, the sea level must rise. But an ocean is a lot more complex than a swimming pool, and we cannot simply multiply the volume of the ocean by the percentage of water expansion and expect to get the right answer. The topic is getting plenty of attention, but the ability to fully quantify this expansion is still some distance away. Recent estimates suggest that from thermal expansion alone, sea level is rising about one millimetre a year.[3]

Measuring sea-level rise is complex. Historically, sea level was measured using tide gauges on land. Unfortunately, the height of the land may change over time, depending on where it is. Along some continental margins, such as north-western North America and India, there is variable up-and-down movement at a rate comparable to the change in sea level – millimetres a year. As we know from the confirmed theory of continental drift (now Plate Tectonics), the continents float, rather like icebergs, on a barely fluid substance known as the Earth's mantle. During the ice ages, Scandinavia and Canada were covered by as much as 3 kilometres of ice. The weight of the ice depressed those continents by several hundred metres. Now that the ice has gone the continents are slowly rising again, so any tide gauges there are slowly being lifted out of the water. Where the land is rising, tide gauges may well report that sea level is falling, regardless of what it is actually doing. In 1991, Bruce Douglas, of the United States National Ocean Service in Maryland, reported a very careful study of the world's tide gauges, using more than 50 years of record.[9] After eliminating all tide gauges on land that were going up or down, he was left with only 21 reliable records. These records showed a highly consistent figure for sea-level change of 1.7 millimetres each year over the 100 years up to 1980.

Since 1992, satellites have been used to measure sea level, and these measurements, unlike those made by tide gauges, are not affected by rising or sinking land and can make determinations of sea level in the open ocean. The Fourth Report of the Intergovernmental Panel on Climate Change used both shore-based gauges and satellite data to reach the same conclusion as Douglas, that sea level had risen during the 20th century

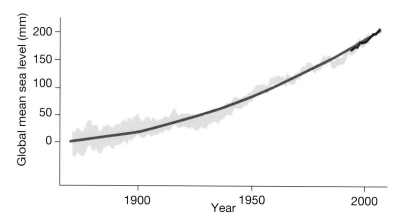

Figure 7.4 Change in global mean sea level since 1870, determined by tide gauges (green band, its width indicates the degree of uncertainty and the darker green line shows the overall trend), and from satellites TOPEX/POSEIDON and Jason1 (red line). Source: after Church et al., 2008.[11]

by about 1.7 millimetres a year but also noted that the rate is increasing. Since 1992, the average rate of sea-level rise has been almost double the 20th-century average, being estimated at 3.3 millimetres per year by Cazanave and colleagues.[10]

The difference between tide-gauge estimates (1.7 millimetres per year) and satellite measurements (3.3 millimetres per year) has led some to query the accuracy of the satellite results. John Church from the CSIRO and an international group of experts on the subject resolved this issue in 2008[11] (Figure 7.4); the rate at which sea level is rising has increased over the past 150 years, so the average rate pre-1980 is inevitably going to be less than the rate after 1980.

Returning to the effect of ocean warming, we saw earlier that the calculated increase in sea level as a result of warming is about 1 millimetre a year, so what is causing the other 2 millimetres or so of sea-level rise? Melting glaciers and icecaps are the other contributors, and they are estimated to be producing a rise of another 1.8 millimetres a year.[12] That leaves at least 0.45 millimetres a year unaccounted for and, basically, this is one of the current unknowns in climate science. (The melting of Arctic sea ice cannot be added because it was already floating in the ocean.)

Sea level is affected by atmospheric change, particularly in barometric pressure, and in Australia particularly by the ENSO. Australia has 16 stable tide gauges, which have been monitored carefully since 1991 by the Australian Bureau of Meteorology's National Tidal Centre. Broadly, sea level is lower during an *El Niño* event and higher during *La Niña*.

These variations, which occur on a time-scale of three to five years, are of the order of 100 millimetres. With these variations averaged out, the overall change in sea level around Australia for the past 20 years is about an 8-millimetre rise on the northern and western coasts, and a 1–2 millimetre rise on the south-eastern coastline. This change is considerably less than the short-term variation caused by weather factors, prompting the National Tidal Centre to state in June 2009 that: 'Caution must be exercised in interpreting the "short-term" relative sea level trends as they are based on short records in climate terms and are still undergoing large year-to-year changes.'

ACID IN THE OCEAN

There is concern that any increase in the level of CO_2 in the oceans, which at present is some 40 times the amount in the atmosphere, will affect ocean life. In Chapter 4 we looked at the balance that the Earth maintains between CO_2 in the atmosphere and in the ocean. Later on in Chapter 10 we will look at the way CO_2 moves between land, sea and air and we will see that as the level of atmospheric CO_2 has risen, approximately a quarter of it has ended up dissolved in the ocean, where it has caused the ocean to become more acidic (see box for background on acids, alkalis and ocean chemistry).

OCEAN CHEMISTRY

Pure distilled water has more than just H_2O molecules in it. A very small proportion of the molecules are split into two parts: a hydrogen atom, which has lost its electron so it is positively charged (H^+) and is called a 'hydrogen ion', and an oxygen atom, which has held onto the other hydrogen so the pair are negatively charged (($OH)^-$) and is called a 'hydroxyl ion'. Only one water molecule in 555 million is split like this. The presence of more hydrogen ions than this makes an acid, while the presence of more hydroxyl ions make an alkali.

Carbon dioxide dissolves in water, and when it does it makes the water weakly acid and also produces bicarbonate ions:

$$CO_2 + H_2O = HCO_3^- + H^+$$

(Baking soda is sodium bicarbonate; dissolve that in water and you get an alkaline solution because many negative bicarbonate ions are produced but no balancing hydrogen ions; the charge is balanced by sodium ions, which do not cause acidity.)

Acidity and its opposite, alkalinity, are measured on a 14-point scale, the pH scale. When chemists work out weights and numbers they use a scale based on Avogrado's number, the number of molecules in 22.4 litres of gas. The number is a constant regardless of the gas, and equals 6×10^{23}. The weight depends on the atomic weight of the atoms that make up the gas, and this weight is called a Mole. 22.4 litres of hydrogen weighs 1 gram; so a Mole of hydrogen weighs one gram. A Mole of oxygen weighs 16 grams, a Mole of water vapour weighs 18 grams, but when it condenses it only occupies 18 millilitres. So one litre of water (1000 grams) contains 1000/18 Moles of water = 55.5 Moles. One hydrogen ion in 555 million water molecules, by weight, then works out to be one 10-millionth of a gram of hydrogen ions in a litre of water, which because a Mole of hydrogen weighs 1 gram, is one ten-millionth of a Mole of H^+. This is how acidity is measured, as the number of Moles of H^+ in a litre. Pure water has a hydrogen ion concentration of 10^{-7} Moles per litre. To simplify the expression, chemists use the negative logarithm of the hydrogen ion concentration; that is, they use just the number 7 for pure distilled water and call it the 'pH', meaning 'weight of hydrogen' (in Latin). pH 7 then means 10^{-7} Moles per litre of H^+.

Battery acid (sulfuric acid) and muriatic acid or spirits of salt (hydrochloric acid) such as is sold to clean mortar from fresh brickwork are highly acidic; their pH value is 0; following from the previous paragraph, this means there is 10^0 Moles of H^+ per litre – that is to say, one Mole per litre (in mathematics $10^0=1$). Caustic soda, used in cleaning drains and ovens, has a pH of 14 (no significant H^+ but one Mole per litre of $(OH)^-$). (Alkaline solutions are also called 'basic' solutions, but as this word has other meanings I will not use it.) Between highly acidic and highly alkaline comes lemon juice and vinegar at 2, beer at about 4, pure distilled water at 7.0, sea water at 8.2, soaps at around 10 and bleach at 12 or more.

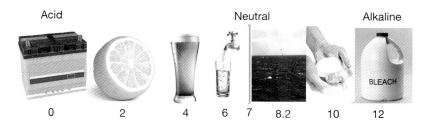

Figure 7.5 The acid–alkaline (pH) range of some common substances. Source (all): Shutterstock com.

The pH of rainwater is slightly acid, being as low as 4.8 in tropical rain, but is more commonly around 5.6. By contrast, the pH of seawater varies between 7.9 and 8.2, which means the oceans are weakly alkaline. The reason for the difference between rainwater and seawater is the presence of calcium in the ocean and calcium carbonate in shells and in ocean sediment. Calcium makes water alkaline, and soils, too, which is why you put lime (crushed limestone) on the garden if the soil becomes too acid. Calcium reaches the sea in rivers, having been dissolved out of rocks as they weathered.

In the sea calcium is used by marine organisms for their shells, and an equilibrium, or balance, is achieved between calcium and dissolved CO_2. Increasing the CO_2 content of the ocean means the ocean becomes slightly less alkaline, or if you like, slightly more acidic. A new balance has to be set up between the CO_2 and the calcium, and this is achieved by the formation of calcium carbonate, which settles to the sea bed and eventually becomes limestone.

$$2CO_2 + 2H_2O = 2HCO_3^- + 2H^+$$

(carbon dioxide and water make carbonic acid – split as bicarbonate and hydrogen ions)

$$Ca^{2+} + 2HCO_3^- \rightarrow CaCO_3 + CO_2 + H_2O$$

(calcium ions and bicarbonate ions make calcium carbonate, carbon dioxide and water)

$$CO_2 + H_2O = HCO_3^- + H^+$$

(carbon dioxide and water make carbonic acid)

So one of the two new CO_2 molecules goes out of the water to form limestone, while the other remains as carbonic acid (a bicarbonate ion plus a hydrogen ion); that is, combining the three equations above: Ca^{2+} (soils) + $2CO_2$ (atmosphere) + $2H_2O$ (ocean) = $CaCO_3$ (shells) + HCO_3^- (ocean) + $3H^+$ (ocean). The ocean has become more acidic.

....

According to the IPCC report of 2007,[13] the oceans have become less alkaline by 0.1 pH units since 1750, currently standing at about 8.08 on the pH scale. Careful measurement of atmospheric CO_2 and ocean pH confirms that the increase in acidity precisely fits the theoretical increase calculated from the CO_2 measurements, so there is no doubt as to the cause of falling pH. These measurements are accurate enough to show the seasonal variation in pH as it follows the seasonal changes in atmospheric CO_2. At the University of Hawaii, marine scientists have been monitoring ocean properties since 1988, and in that part of the Pacific Ocean have found a steady increase in acidity (lowering of the pH) amounting to 0.04 pH units over 20 years.[14] It took 250 years for ocean pH to drop by 0.1 units; at current rates that increase in acidity will take only 50 years. A change from 8.1 to 8.0 may not seem like much, but like the Richter scale for earthquakes, a change of one unit means a 10-fold change in impact. A pH change of 0.04 is a 10 per cent change in acidity.

From China comes a study of corals in the South China Sea.[15] From 7000 years ago until 1200 years ago, the pH record obtained from the corals showed a gradual increase, from 7.95 to 8.15. The current value measured in this work is 7.85, indicating a bigger decline in pH over the past 1000 years than the gradual increase of the previous 6000 years. A decline was also found in a study of corals in the Great Barrier Reef;[16] from 1800 to 1940 average pH has dropped from 8.0 to 7.8.

Ocean scientists recognise that the consequences of changes in pH on marine organisms are poorly known. Some have even voiced concern that corals and other sea creatures will die as a result. The pH of the water in a coral lagoon of the Coral Sea has been found to vary in accordance with a large-scale, 20–30-year variation in ocean temperature, called the Pacific Decadal Oscillation.[17] This phenomenon causes, among other things, a variation in the tropical Pacific sea-surface temperature of about half a degree and a systematic variation in pH from 7.9 to 8.2. The authors of the study pointed out that the corals they examined were adapted to this range in pH. At present the cycle is at a high pH. By 2035, if the ocean pH continues to decline as it is at present, the authors concluded: 'the extent to which corals and other calcifying reef organisms can adapt to such rapid decreases in pH is largely unknown'.

In the journal *Science* in 2007, an international group of 17 scientists[18] wrote that experimental studies showed that a doubling of atmospheric CO_2 from pre-industrial levels has led to a 40 per cent decrease in coral growth by inhibiting aragonite formation (the calcium carbonate mineral of corals). They coupled this experimental evidence with the geological record that shows an absence of coral reefs in marine sediments of 200 million years ago, when CO_2 levels were much higher than they are today. From this they concluded that if CO_2 levels continue to increase as they have until now, by the end of the 21st century most coral reefs will have undergone very significant structural change through the loss of many reef-building species.

At about the same time, another group led by Ilsa Kuffner[19] published experimental findings of the growth rates of certain calcifying algae that encrust coral reefs. The authors set up growth chambers under two different atmospheric CO_2 levels: 400 ppm and 765 ppm. They found significant pH difference between the two setups and a much lower growth of the algal crusts under the higher CO_2 regime.

Another study[20] found that if the seawater pH were reduced to about 7.5 from its normal 8.1, corals gradually lost their limestone skeletons but did not die. After a year, when the pH was restored to 8.1, they rebuilt their skeletons.

Corals and associated encrusting algae may decline or vanish, but tiny marine plankton of a particular type called coccolithophores, such as shown in Figure 7.6, may do better under increased CO_2. An international research team under the leadership of Iglesias-Rodriguez and Halloran[21] concluded that there was a significant increase in the size of the shells of one species of these tiny creatures as the atmospheric CO_2 levels increased, whereas an earlier study by Riebesell and colleagues of the same plankton had found the reverse.[22]

Figure 7.6 A coccolithophore, *Emiliania huxleyi*, grown under CO_2 levels twice that of today. This creature is about 6 micrometres across (a human hair is more than three times this width). Source: Paul Halloran.

Some months later, Riebesell's study group refuted the Iglesias-Rodriguez group's result, arguing that there were flaws in the experiment.[23] As is normal, polite, scientific practice, Iglesias-Rodriguez and her colleagues were offered the opportunity to rebut the criticisms; they explained their work further and in turn criticised the experimental setup of Riebesell et al.[24] In the meantime, four of the Iglesias-Rodriguez and Halloran group scientists emphasised one aspect of their original paper, namely that in sediments deposited between 1750 and the present, the size of the larger plankton had increased.[25] The debate continues and, importantly, it will eventually be resolved because disagreements such as this are how science progresses.

Around Antarctica, the ocean supports huge stocks of creatures, ranging from whales to their tiny food, krill. One important group of tiny organisms are the pteropods. They float in the sea protected by a thin shell of calcium carbonate made from the mineral aragonite, the same mineral as corals use for their structures. Aragonite dissolves in slightly less acidic seawater than does its chemically identical cousin calcite, the component of almost all limestone and many seashells. Both kinds of shell are more soluble at depth than near the surface, so deep water today is like shallow water will be in 30 years if CO_2 levels continue to rise. There is a line known as the 'aragonite saturation horizon' that separates waters in which pteropods can happily make their shells from waters in which they cannot. At present it is between 1000 and 2000 metres depth in the sub-Antarctic (47°S) Southern Ocean. By 2030, when CO_2 reaches 450 ppm, the line will be at the surface for waters south of the polar front, and soon after for sub-Antarctic waters. Figure 7.7 compares two shells, one sampled at 1000 metres and the other at 2000 metres.[26] The picture tells it all. This organism is on its way out, and with it will go all those larger creatures that depend on it for food.

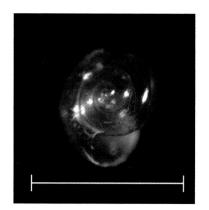 

Figure 7.7 Pteropod shells taken at 1000-metre (left) and 2000 metre (right) depths from the Southern Ocean. The dissolution of the aragonite shell taken from deeper water is very obvious. Scale bar is 1 millimetre. Source: Donna Roberts.

These are just a few specific examples of the research that warns us of the impact increasing atmospheric CO_2 is having on the world's oceans. A review of ocean acidification published in 2009[27] summarised much research into the impacts of changing pH and ocean CO_2 content on marine organisms. The authors recognised that 'The potential for marine organisms to adapt to increasing CO_2 and the broader implications for ocean ecosystems are not well known; an emerging body of evidence suggests that the impact of rising CO_2 on marine biota will be more varied than previously thought, with both ecological winners and losers.' However, they concluded that 'Acidification will directly impact a wide range of marine organisms that build shells from calcium carbonate, from planktonic coccolithophores and pteropods and other molluscs, to echinoderms, corals, and coralline algae' and further, that 'Acidification impacts processes so fundamental to the overall structure and function of marine ecosystems that any significant changes could have far-reaching consequences for the oceans of the future and the millions of people that depend on its food and other resources for their livelihoods.'[27]

OCEAN TEMPERATURE

There is another quite different consequence of changing climate for marine life, and that is the effect of changing water temperature. Most organisms have their preferred zone of temperature range, and if that zone shifts north or south in response to oceanic temperature change, so the organisms will go with it. From microscopic algae to whales, a shift in one animal will be followed by a shift in the ones that eat it. In 2006, the CSIRO prepared a report to the Australian Government on the known and predicted effects of climate change on Australian marine life. For most groups of organisms, the conclusion was that there was so far no known effect, largely because of a lack of research on the topic. Some species of fish used in commercial fishing have expanded their range southwards, and this was concluded to be a response to warming of the ocean. For some other more abundant types, including tuna, sharks, sardines and squid, there is so far insufficient knowledge. A prediction for most marine species of the southern hemisphere is a movement of their range southward.

It is all very well for fish to migrate around the ocean to remain within their natural temperature range, but animals that live on the ocean floor do not have that luxury. Probably the most closely studied ocean creatures of this kind are corals. Corals such as those of the Great Barrier Reef are a combination of two organisms, an animal and a plant, living in what is termed 'mutualistic symbiosis'. The two need each other for survival. In the case of corals, the animal is called a 'polyp', and lives inside a tubular shell made of the mineral aragonite. Inside and among the polyp and its tentacles live algae, which are microscopic plants. The plants take in CO_2 from the polyp as well as essential nitrogen; the polyp gets oxygen from the algae as well as nutrients such as glucose. Together, they survive; apart they do not. It is the algae that gives coral its colour, and these algae are highly sensitive to changes in temperature; if the ocean's temperature gets too warm the algae are lost or their number greatly declines and the coral loses its colour – it bleaches (see Figure 7.8, p. 120). As long as the temperature does not remain high for too long, the algae will return and the coral is restored.

There have been three relatively recent major bleaching events in the Great Barrier Reef. Worldwide, bleaching occurred in the summer of 1998–9, with corals in some habitats suffering 95 per cent mortality. In the Great Barrier Reef, two-thirds of the inshore reefs were bleached, though only 15 per cent of the offshore reefs were affected and within a year most of the reefs had recovered. In 2002, about 60 per cent of the reef was affected to some extent, but again almost all recovered.[28] A more localised bleaching occurred in 2005–6 around the Keppel Islands, with 40 per cent mortality,[29] but the reef, which is made up mainly of fast-growing corals, bounced back within a year.[30] Each of these bleaching events happened during a 2°–3°C increase in ocean temperature.

Coral reef research clearly shows that while a few days at 29°C causes bleaching in the southern part of the reef, northern reefs thrive at that temperature but would be damaged at 31°C.[31] This suggests that corals

Figure 7.8 Bleached (left) and healthy (right) coral from the Keppel Islands after the 2002–3 bleaching event. Source: Ove Hoegh-Guldberg, Global Change Institute, University of Queensland.

and their algae are adapted to survive in their own particular place. What is not known is how fast coral–algal evolution can happen. Could rapidly increasing ocean temperatures allow enough time for them to adapt to the warmer water? If they cannot change quickly enough, such shallow-water corals are doomed by global warming.

SUMMARY

Topic	Observation	Key statistic	Conclusion
Ocean heat content	The Earth's major heat store, will maintain global temperatures for centuries.	Steady rise since 1950.	Ocean is expanding as it heats, and so sea level is rising.
Sea level	Has been rising for the past 100 years.	Rose 170 mm from 1880 to 1980, now rising at more than 3 mm/year.	Combination of warming ocean and melting ice are raising sea levels.
Ocean acidity	Increasing.	10% more acid over 20 years.	Increased CO_2 decreases ocean pH, impacts on many marine organisms.
Sea-surface temperature	Rising.	~1°C in the past century.	Causes coral bleaching and death, shifts ocean creatures' habitats.

There are a number of other factors that might be assessed to examine the question of global climate change. In the forefront are changes in faunal distribution, and Professor Tim Flannery's book *The Weather Makers* most elegantly explains these matters. The evidence for climate change is clearly before us, and we can now answer the first question more completely: Is the climate changing?

Answer: Yes, it is changing. It is getting hotter, rainfall distribution is changing, ice is melting, the ocean is becoming more acidic and the sea level is rising.

In chapters 3 and 4 we also saw the answer to the second question: What can change the climate? That leaves the third question: How has climate changed in the past?

Some authors point out that there have been many climate changes in Earth's distant past, and that today's climate change is nothing special. To look at that point of view, and to answer the third question, the next two chapters will review 700 million years of the Earth's history.

FURTHER READING

Church J, Woodworth P, Aarup T & Wilson S (2010) *Understanding Sea-level Rise and Variability*. Wiley–Blackwell.

Zeebe RE (2012) 'History of seawater carbonate chemistry, atmospheric CO_2, and ocean acidification'. *Annual Review of Earth and Planetary Sciences*, *40*, 141–65.

8

FROM ICE-HOUSE TO GREENHOUSE

What's past is prologue.

Shakespeare, *The Tempest*

The world is a frozen ball. From the poles to the equator there is nothing but ice. The mountains are covered by snow, the valleys are filled by glaciers and across the plains, ice sheets are kilometres thick. From the poles to the equator there is no water. Ice covers the oceans, barely breaking into floes across the tropics. It never rains. Where the sun is hottest a little water vapour sublimes from the ice, eventually to fall as snow. The whole Earth is silent.

But not still. Beneath the ocean, erupting volcanoes pour lava onto the sea floor and gases into the cold water. The gases remain there, dissolved; kept in by the blanket of ice. On land, too, volcanoes erupt. Out of the rifts pours carbon dioxide as well as molten rock. Like the gases in the ocean, the CO_2 in the atmosphere stays there, for it has nowhere to go. There are no plants to take it up, no open ocean to absorb the CO_2, and no rocks are exposed to weathering. Slowly, the levels of CO_2 in the air rise, and slowly the world warms, until at last some of the sea-ice melts. Now, the CO_2 in the oceans also starts to escape into the air, and as it does the opening ocean absorbs ever more of the Sun's heat. Together, the increased CO_2 and the darker surface of the seawater combine in a feedback loop that, after perhaps 10 million years, releases the icy grip of 'snowball' Earth.

Open to argument, certainly, but there are many geologists who suggest that something along those lines describes the world before the Cambrian period, 700 million years ago. The idea was conceived by Joe Kirschvink at the California Institute of Technology in the United States and published in 1992.[1] Since then, geologists have uncovered much supporting evidence, extensively summarised by Paul Hoffman and Daniel Schrag from Harvard University.[2] By the time the climate had stabilised again there was no ice, the atmosphere had perhaps 6000 parts per million (ppm) CO_2, which is some 15 times current levels,[3] the temperature was about 8°C warmer than today's[4] and lichens and mosses had started to colonise the land. Fed by the high CO_2 levels, plants evolved to ferns and small trees, and by the Devonian period 200 million years later, the land was green. The spreading of plants had a major effect on climate; they accelerated rock weathering, drawing CO_2 down into the soil as calcium carbonate, and they absorbed CO_2 themselves. Both processes would have reduced the temperature, but according to Guilliame Le Hir at the University of Paris 7, as plants spread across the bare Earth, they reduced the Sun's reflection, allowing more heat to be absorbed.[4] As a consequence, the temperature would not have fallen quickly.

Figure 8.1 Left to right: plant evolution, from mosses to ferns, to seed-bearing but non-flowering plants such as the *Gingko biloba* and to flowering plants.

There are many ways scientists work out climates of times long past. As valley glaciers grind their way down the valley they break up the rocks beneath them. Bits of rock, sand and silt become incorporated in the ice. When the glacier reaches the sea (if it does), it cracks and large masses of ice fall off to become icebergs. They float over the ocean, slowly melting and dropping their contained load of rock onto the sea floor. Ages later, that sea-floor sediment, compacted and turned to rock, may be forced up into mountains by the processes that shift the continents. When geologists looking at those rocks recognise extensive deposits of ice-rafted 'drop stones' of the same age in many different parts of the world, they are able to conclude that the era was cold. Another method uses the isotopic composition (see Isotope box in Chapter 5) of the seashells in the ocean sediment and so determines the temperature of the water in which they lived.

Evidence for the atmospheric CO_2 levels of the geologic past also comes from a variety of studies. One involves the carbon in long-buried soils. Soil is a wonderfully complex substance. It is a mixture of rock and mineral fragments, clay, water and air, which provides a home for a myriad of animals such as ants, worms, beetles and bacteria, as well as fungi and innumerable plant roots. The air in the soil is richer in CO_2 than the atmosphere above it, thanks to the activities of all those soil creatures. And because of the higher CO_2 levels, soil water is more acidic than stream water, typically having a pH of about 4.5. The water and air in the soil slowly weather the minerals and rocks and, in so doing, particularly in drier climates, some of the CO_2 is converted to particles of limestone (calcium carbonate). By analysing the amounts of carbonate in ancient buried soils, geologists can work out how much CO_2 would have been in the air of that past time.

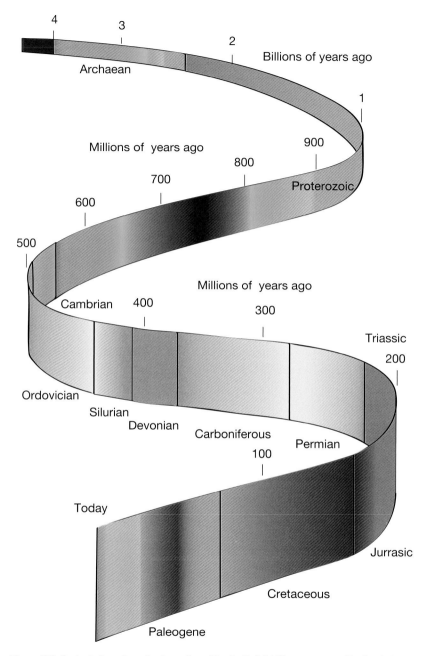

Figure 8.2 Geologic time, from the formation of the Earth 4.6 billion years ago. Shading indicates the climate of the time: redder is warmer, bluer is colder.

Detail in the structure of fossilised leaves provides another and quite independent way to estimate past CO_2 levels, and study of the composition of the shells of tiny ocean creatures, called 'foraminifera', also yields an analytical tool. Coal deposits give another clue to ancient climate. Coal is the fossilised remains of plants; extensive deposits of coal suggest conditions in which there was plenty of CO_2 in the air, because plants need CO_2 to grow.

THE PAST 400 MILLION YEARS

Putting the geological clues together, as has been done by various researchers, paints a picture such as shown in Figure 8.3. Four hundred million years ago it was warm and the CO_2 levels in the atmosphere were 10 times those of today. This encouraged plant growth and led to the major coal deposits of the world. But the plants took the CO_2 out of the atmosphere, and as they did the temperature fell, leading to extensive polar ice sheets. Later the temperature warmed, the CO_2 levels rose and more coal deposits were laid down. Then, again, the plants took the CO_2 out of the atmosphere and the temperature fell, ending in the recent ice ages.

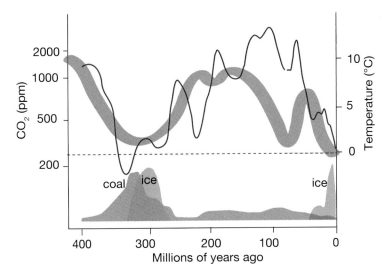

Figure 8.3 The relationship of climate evidence and deduced CO_2 levels in the atmosphere for the past 400 million years. Estimated temperature is the upper red line, from Zachos et al. (2001)[6] for the period 0–65 million years ago and after Beerling et al. (2009)[5] for the period from 65 million years ago to 400 million years ago. CO_2 through the ages is shown as the broad grey band, after Park & Royer (2011).[9] Temperature and CO_2 have changed roughly in step, with ice ages and coal deposits providing some of the evidence.

From such lines of evidence the changes in temperature and CO_2 and the eras of glaciation can be mapped out over the past 400 million-year period. The connection between climate and CO_2 is quite evident in the geological record, and that evidence – the coal, glacial drop-stones, types of shellfish fossils – tells us that when atmospheric CO_2 falls, so does the temperature; when CO_2 rises, so does the temperature.

The younger the rocks, the better they preserve the evidence of past climates, and so it is that a quite detailed knowledge of global temperature of the past 65 million years has been worked out from studies of the shells of various small deep-sea creatures. James Zachos from the University of California in Santa Cruz, United States, is the lead author of a summary of the Earth's climate since the extinction of dinosaurs (Figure 8.4).[6] Starting from a temperature about 8°C warmer than today's, the temperature climbed another 4°C until 50 million years ago. Along the way there was a sharp spike in ocean temperature, amounting to about 6°C over 10 thousand years, with an associated CO_2 spike, falling back again over the following 200 thousand years. This event is known as the Paleocene–Eocene Thermal Maximum, and may well have been caused by a burst of methane.[7] One consequence was the extinction of a major group of deep-sea organisms and of shallow-water reef builders. The effects of increased temperature, dissolved CO_2 and increased acidity cannot be separated, but it seems likely there was a very sharp drop in ocean pH, though much more slowly than today's unparalleled rate of decrease.[8] This is probably the most rapid global temperature rise in the geological record, but it happened at a rate 20 times slower than the rise in global temperatures during the 20th century. After that there was a slow but irregular decline into the ice ages.

Much of our knowledge of the role of CO_2 in geological history comes from the Yale University laboratories of Bob Berner. Professor Berner is one of the most highly honoured geochemists of the past 50 years. Many of the conclusions from which figure 8.3 is drawn come from his work and that of his colleagues[9]. These scientists regard greenhouse gases, particularly CO_2 and methane, as the main drivers of climate change over the geological past, with variations in the Sun considered an additional factor.

Taking a different view are two scientists, Nir Shaviv and Jan Veizer, who have determined temperatures from analyses of the composition of seashells, and have concluded that it is entirely possible that the dominant cause – they say about two-thirds of long-term climate variation is celestial – results, they claim, from variation in the Sun's output and in the

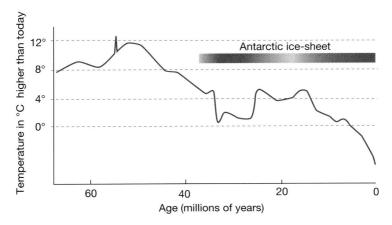

Figure 8.4 Ocean temperature over the past 65 million years, after Zachos et al.[6] The extent of the Antarctic ice sheet, which developed about 37 million years ago, is shown by the grey and black shaded bar.

cosmic rays that reach the Earth from the galaxy.[10] The remaining third might be driven by CO_2 variation.

Climate changes of the past were slow, as so many geological processes are, though sometimes there were fast changes. Most of those are called 'catastrophes', and they would have been such for the organisms then living, including the meteorite impact that killed off most of the dinosaurs. But from a mammal's point of view the meteorite was a blessing, because these warm-blooded creatures were able to survive the years of cold caused by the meteorite's dust.

Comparing our own climate to those far, far distant times is in one way pointless. Hot or cold, wet or dry, poisonous or fragrant, any changes in the atmosphere and climate were such as to allow life to evolve to what it now is. Back in the age of dinosaurs, before they were extinguished by that meteorite, from 100 to 200 million years ago global temperatures were as much as 10°C warmer than today. Now, that might well have suited a cold-blooded creature like a dinosaur, but heat waves reaching above 50°C would not suit us humans. For just one example, there was a severe heat wave in England in 2003. The average summer temperature in southern England is 21°C. For 10 days in early August that year, the maximum temperature was over 32°C, with a record maximum of 38.5°C. During that period 2139 more people died than would normally die, the majority on the days of extreme heat. A heat wave in the Cretaceous period would have been much hotter than that – probably around 50°C.

To say, as some sceptics do, that the world can cope with 10 times as much CO_2 as we now have because it did so in the Cretaceous period 100 million years ago ignores a host of issues. Humans were not around then, and the animals that were had evolved to live with that level of CO_2. Then, as it slowly declined, so the organisms adjusted – or died.

THE PAST MILLION YEARS

The climate story of long ago is still the subject of energetic research and discussion. The evidence is difficult to gather, buried as it is in the rocks. As we look ever closer to the present, the task of deciphering the climate gets easier because the evidence is more abundant and more easily interpreted. We therefore have a very good idea about what has happened over the past million years. During that period the northern hemisphere, in particular, and the southern hemisphere perhaps less dramatically, were covered by continental ice sheets and then uncovered, eight times in all. The reason for these ice ages has been well explained by variations in the spatial and seasonal patterns of the Sun's energy reaching the Earth, the Milanković cycles.

MILANKOVIĆ CYCLES

We saw briefly in Chapter 3 that the ice ages were triggered by periodic variations in the Earth's orbit around the Sun. Figure 8.5 shows the remarkable correlation over the past 140 000 years between the amount of the Sun's energy reaching the northern hemisphere and the global temperature.

The initial evidence for glaciation was in the recognition of ice-deposited boulders and sediment over country that is now rich agricultural land, such as southern Canada and the northern United States, Scandinavia and northern Europe. New York was covered by a kilometre of ice 15 000 years ago and northern Canada by 3.5 kilometres of ice. That ice melted as a result of global warming, over a period of about 10 000 years.

ICE CORES

Carefully drilled and extracted cylinders of ice – ice cores – have proven to be one of the most valuable proxies, not only for temperature but also

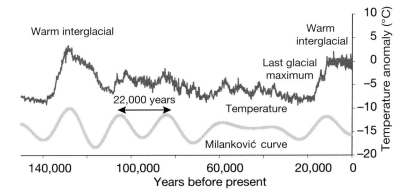

Figure 8.5 Relationship between the Milanković curve and temperature. The lower wavy line, labelled Milanković curve, shows how the amount of sunshine reaching the northern hemisphere at 65°N has varied over the past 150 000 years. The upper curve is the temperature anomaly determined at Vostok in Antarctica, based on hydrogen and oxygen isotope measurements.

for the content of CO_2 and other gases in the atmosphere. Polar snow quickly compacts to ice, and variations in the snow's composition show up as annual layers. Added to that are occasional deposits of volcanic dust, and each of these can be related to a known volcanic eruption. Simply by counting annual layers, the age of any part of the ice core can be found and, as they are counted, the accuracy of the count can be checked against the volcanic dust. As well, there are methods using the amount of methane trapped in the ice and the isotopic composition of the ice itself, which provide checks on the age of any particular layer.[11]

Trapped in the frozen snow are bubbles of air, and from these the scientists can measure the composition of the atmosphere at the time of the snowfall, especially the CO_2 and methane content. And, perhaps most importantly, they can find the temperature of the snow when it fell, by analysing the hydrogen and oxygen isotope composition of the ice and its bubbles of gas. So it is that the ice cores yield precious information about past climates.

With its thick ice cap, Antarctica is an ideal place to sample ice from ages past. Cores have been taken there from the late 1960s, with the core drilled at the Vostok Research Station in 1996 being one of the most studied.[12] This core went to a depth of 3600 metres and gave an unprecedented picture of temperature change over the past 400 000 years. The multinational European Project for Ice Coring in Antarctica (EPICA) later drilled an ice core that gave climate evidence for the past 740 000 years. Its results closely match those from the Vostok core, and show an additional four ice ages.[13] A third, deep Antarctic core drilled at the Japanese research station Dome Fuji gives records going back 720 000 years. A short article by Hideaki Motoyama described how such ice cores are drilled.[14]

During the depths of the ice ages the temperatures in Antarctica were generally 8°C lower than today, and during the interglacials about 3–4°C warmer[15] (Figure 8.6) and possibly higher.[16] This temperature range does not apply to the entire globe. The polar temperature range from ice age to interglacial is much greater than the global variation.[17] For example, since 1979, Arctic surface temperatures have increased twice as fast as the northern hemisphere average.[18] The difference, called the polar amplification factor, is thought to be around two times, thus an Antarctic temperature range of 12°C would be equivalent to a global variation of 6°C.[19]

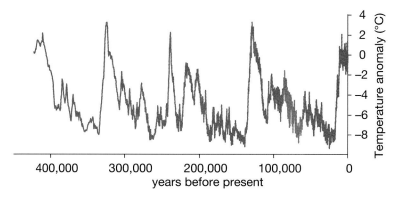

Figure 8.6 Temperatures deduced from the Vostok ice core; values are relative to today's mean global temperature. Source: Petit et al., 2001.[20]

Temperature estimates for the more recent past were published in 2006 by a group of nine scientists from four US universities who had made a detailed study of glacial ice cores from the South American Andes.[21] One core collected data from almost 12 000 years ago, and revealed from the variation in oxygen isotope ratios that the region was probably 1.5 to 2°C warmer than present during a period 8000–9000 years ago following the end of the last ice age (see Figure 8.7). They compared this record to a similar one from Mt Kilimanjaro and to the change in solar radiation deduced from astronomical data (Milanković). They concluded that the temperature changes found in their ice-core data were largely driven by slow changes in the Sun's radiation in the tropical northern hemisphere. A very similar result was obtained from a study of northern European

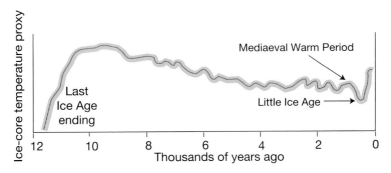

Figure 8.7 The temperature variation as determined by tropical glacier ice-core proxy over the past 12 000 years. Source: after Thompson et al., 2006. This is, in effect, a close-up of the short section at the extreme right of Figure 8.6.

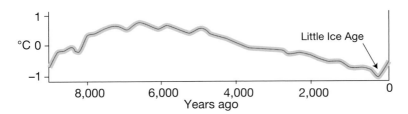

Figure 8.8 Temperature variation from the mean over the past 9000 years in northern Europe, based on pollen analysis. Source: after Seppä et al., 2009.[22]

pollen (Figure 8.8).[22] The authors noted that the past 150 years have shown the strongest warming trend since 8000 years ago.

Compare the past 1000 years in Figures 8.7 and 8.8 with that in Figure 1.2, Michael Mann's 'Hockey Stick'. These research works, published long after the original Hockey Stick, shows exactly the same features: slow cooling until the end of the Little Ice Age, about 1800, and then a sharp, continuous rise.

It is important to appreciate the rate at which the Earth's natural temperature changes occur. From the warm interglacials to the depths of each ice age takes about 20 000 years, and the world cooled by 5°–6°C each time; a rate of about 1°C in 4000 years. Warming up after each ice age was quicker: 5°C in 6000 years or 1°C in 1000 years. The last warming ended 10 000 years ago, and since then the world started to cool again, quite slowly this time, at about 2°C in 10 000 years. At present the world is warming at the rate of 1°C in 60 years; that is, 20 times faster than any previous sustained rate of temperature change.

ABRUPT CLIMATE CHANGE

The rate of change deduced from global temperature estimates is one thing; local variation is entirely another. We saw in earlier chapters that local experience may suggest major climate change, whereas global averages reveal only minor trends. Such an example is that of the Sahel, where severe drought has been continuous in a world in which total rainfall has, if anything, slightly increased. In the same way, temperature and climate changes as fast as several degrees in 3 years have been recorded in ice cores. As the world emerged from the last ice age, the records reveal huge local to regional temperature fluctuations. The most pronounced of these marked the end of the last glaciation, when sea level started to rise very rapidly as the result of ice caps melting. A highly significant event, mostly felt in the northern hemisphere, is known as the Younger Dryas, a time when there was a return to cold conditions lasting from 12 800 to 11 500 years ago (Figure 8.9). In Greenland (lower curve in Figure 8.9), pulses of cooling associated with periodic outbursts of melt water led into the Younger Dryas, whereas in Antarctica (upper curve) the changes were far less pronounced.

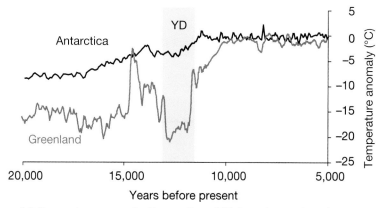

Figure 8.9 Temperatures across the Younger Dryas (YD). Large temperature changes are evident over Greenland (lower curve), but a much smaller and more gradual change is revealed from Antarctic ice cores (upper curve). Sources: Antarctic data from Petit et al.,[20] Greenland data from the GISP2 ice core.[23]

The end of the Younger Dryas was marked by an even faster warming, but the rate of the temperature change varied from place to place (Table 8.1). Clearly, the warming as the world came out of the last glaciation was not globally uniform. Rapid changes in Greenland and northern Europe

Table 8.1 Temperature change at the end of the Younger Dryas

Place	Temperature rise °C	Duration (years)	Rate (°C/year)	Source[24]
Greenland	5–10	50	0.1	Dansgaard et al. (1989), Severinghaus et al. (1998), Alley (2000)
Spain	7	40	0.1	Rodrigues et al. (2010)
Switzerland	4.5	200	0.02	Blaga et al. (2010)
Indonesia	3	1000	0.003	Griffiths et al. (2010)
south-western Pacific region	6	4000	0.0015	Gagan et al. (2010)
Australia	0	1300	0	Calvo et al. (2007) Tibby (2012)
Antarctica	3	1300	0.002	Vostok ice-core data[20]

have not been found for the southern hemisphere, and in Australia there is no record of the Younger Dryas cold interval.

What lay behind these changes (see the references in Table 8.1 for details) was the break-up of the northern hemisphere ice sheets, with sporadic collapses delivering huge quantities of freshwater to the northern Atlantic.[25] This, in turn, interfered with the Great Ocean Conveyor Belt in such a way that the southern oceans warmed more rapidly, releasing CO_2, which in turn accelerated global warming.[26] Rapid melting of north Atlantic sea ice eventually allowed warmer water to the surface, initiating the rapid warming across Greenland and northern Europe.

A smaller, abrupt event happened 8200 years ago, when the temperature dropped by 10°C in the north Atlantic, and by 1–2°C across the northern hemisphere.[27] In Greenland, where the most detailed records available come from ice cores, the temperature dropped 3°C in 20 years, then after 60 cold years, it took another 70 years for the temperature to return to its original level. In this case, as the great North American ice sheet melted, a huge lake, known as Lake Agassiz, formed.[28] Suddenly, the dam holding back the melt water burst, and a vast volume of water cascaded into the north Atlantic, instantly lowering the ocean temperature and, with it, the surrounding lands. Such local catastrophes are thought to explain the many such fluctuations in temperature recorded at different times across the northern hemisphere.

Less severe, but still, from a human point of view rapid climate changes are recorded over the following 5000 years across the northern hemisphere. These events show evidence of cooler poles, such as alpine glacier advancement, lower temperatures of the order of 1°C recorded in ice cores, and drier tropics.[29]

It is important to understand that climate change, whether it occurring many millions of years ago, throughout the ice ages, or during the 20th century, does not always have to have had the same driver. The attribution of 20th and 21st-century climate change to increasing CO_2 in the atmosphere is but one of the several forces that can change the climate. If Shaviv and Veizer[30] are right, perhaps the very long-term changes in climate are forced from outside the Earth, possibly from outside the solar system. Whatever the primary cause, we can be sure that CO_2, water vapour, albedo and all will amplify the effect. The Milanković cycles drove the planet into a cycle of ice ages and interglacials. A burst of methane probably forced the sharp climate change of 55 million years ago.

THE AMPLIFIER

The Milanković cycle changes in the Sun's influence on the northern hemisphere changed the ice-age climate. But the Sun had help. It was helped by changes in surface reflection of the sunlight. Each decrease in the Sun's seasonal warmth increased the amount of snow, which in turn decreased the sunlight absorbed by the Earth, which decreased the temperature. This feedback amplified the cooling. Conversely, at the end of each glaciation, snow melted and so the thermostat ramped up to accelerate the warming.

Greenhouse gases amplified the feedback. We know about the greenhouse gases because trapped in the ice are bubbles of air, from the time when the snow was frozen into ice. Small as these bubbles are, scientists can measure their composition to great accuracy, and have produced a very clear picture of how the atmosphere and temperature have related throughout the 400 000 years of the ice ages.

A closer look at the Vostok ice core shows a remarkably close coherence between the Sun's irradiance on the northern hemisphere, the ice temperature and the atmospheric CO_2 and methane (CH_4).

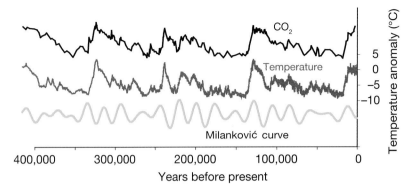

Figure 8.10 Data from the Vostok ice core taken from NOAA. The lowest curve is the variation in the Sun's irradiance at 65° north latitude according to the Milanković cycles. Source: www.climatedata.info. The central curve is the ice-temperature determination and the upper curve is the atmospheric CO_2 content. Source: Petit et al., 2001.[20]

Figure 8.10 shows results from the Vostok core. At the bottom are the variations in the Sun's irradiance reaching the Earth at 65° north latitude, the Milanković curve. Above that is the variation in temperature. You will notice that each peak in the Sun's input, at least for the past 200 000 years, corresponds very closely with a steep part of a temperature increase, though there is less of a match the further back in time you go. The temperature anomaly maximum is never more than 3°C above the baseline zero (defined as the annual average of the present temperature at Vostok, which is minus 55°C; it is the coldest place on Earth). Finally, there are the CO_2 measurements; they have a common minimum value during the ice ages of about 180 ppm, they never exceeded 290 ppm during the warm periods between the glaciations and they follow along with the temperature. Or do they?

Close studies of this and other ice cores show that the CO_2 rise lags behind the temperature rise by a bit less than 1000 years.[31] So here is a strong clue about cause and effect. If the CO_2 rise occurs *after* the start of the temperature rise, the temperature rise cannot be caused by the CO_2 increase.

On the other hand, the temperature rise might have caused the CO_2 rise, and in Chapter 4, in considering the Earth's thermostat we have already seen why. As the ice age developed thanks to orbital changes, the oceans soaked up more of the atmospheric CO_2 than they held when it was warmer, and this is one contributor to the cooling, the less CO_2 in the air, the less the blanketing or greenhouse effect. And the opposite is true. As the Sun's heat increased in the high latitudes of the northern hemisphere, the world started to warm. It can take as long as 1000 years for the surface changes to be fully distributed throughout the oceans (see Chapter 3), and as they slowly warmed they released CO_2. More CO_2, more greenhouse effect, so more warming and more CO_2 is released, and so on.

It is from these records that we are able to be so confident about the importance of feedback on global temperatures. The variation in the patterns of the Sun's energy reaching the top of the atmosphere between ice age and warm interglacial amounted to no more than 0.7 per cent. The variation in the Earth's temperature, about 6°C, is a change of 2 per cent (for such comparisons the temperature has to be measured in degrees Kelvin, not degrees centigrade. The Kelvin scale starts at minus 273°C, so the variation was from about 280 to 286°K). The only way a change of 0.7 per cent in incoming energy can be translated into a warming of 2 per cent is via the amplification caused by greenhouse and reflection feedbacks, explained above and also in Chapter 4.

In 500 million years the Earth has undergone change, not just in climate but also in the composition of the atmosphere. Oxygen has steadily increased from an estimated 2–3 per cent in the Cambrian period to 21 per cent today. CO_2 has fluctuated between 200 and perhaps 6000 parts per million. A sudden burst of methane is thought by some to have triggered the change from snowball to hot-house Earth, 635 million years ago.[32] Another methane burst may have triggered a sudden global warming and the extinction of many marine creatures 55 million years ago. 'A sudden burst of methane' sounds dramatic, until you realise the actual time frame. In the geological record, 'sudden' could mean over half a million years. Vast changes have occurred, and they took vast periods of time.

SUMMARY

Topic	Observation	Key statistic	Conclusion
The geologic past	Temperature and atmospheric CO_2 varied in harmony.	At times, global temperatures were as much as 10°C warmer, millions of years ago.	Climate has always been linked to atmospheric greenhouse gases.

Topic	Observation	Key statistic	Conclusion
The ice ages	Triggered by changes in seasonal and regional patterns of the Sun's radiation.	6° temperature range from glacial to interglacial.	CO_2 is a climate-change amplifier.
Milanković cycles	Explain repeated ice ages.	8 ice ages in 1 million years.	The rate of change was very much slower than today's.
Abrupt climate change	Locally, annual mean temperatures changed rapidly as the ice melted.	3°C change 8200 years ago across Greenland and North America caused by ice-water flood.	Unique, local or regional conditions then, are not present today.

Except for the ice melt, abrupt climate changes of 10 000 years ago caused by the melting ice sheets, we know of nothing in the geologic record that compares with the rate at which the world is now warming. But perhaps we cannot 'read' quick changes in times long past. If we look closer to the present, might we find sudden temperature changes? A realistic perspective on climate change demands a full appreciation of how it was in historically and geologically documented times of the pre-industrial world, and that is the subject of the next chapter.

FURTHER READING

Frakes LA, Francis JE & Syktus JI (1992) *Climate Modes of the Phanerozoic: The History of the Earth's Climate over the Past 600 Million Years*. Cambridge University Press.

Hodgkinson T, Jones MB, Waldren S & Parnell JAN (2011) *Climate Change, Ecology and Systematics*. Cambridge University Press.

9

THE PAST 2000 YEARS

The Thames seems now a solid rock of ice, and booths for the sale of brandy, wine, ale and other exhilarating liquors, have been for sometime fixed thereon; but now it is in a manner like a town: thousands of people cross it, and with wonder view the mountainous heaps of water, that now lie congealed into ice.

Dawk's Newsletter, 14 January 1716

Records of the river Thames in London having frozen go back to as far as 1063, and the event – for it did become an 'event' featuring 'frost fairs' – occurred not infrequently between about 1500 and 1820. After old London Bridge was replaced in 1831, ice no longer accumulated against its supports, allowing the river to flow more freely, and the Thames has not frozen since. The period from 1400 to about 1800 is now called the Little Ice Age.[1]

According to the Milanković curve described in Chapter 3, we should have been experiencing a slow cooling trend for the past several thousand years, and Figure 8.5 shows exactly that. There are gentle variations on that slow trend; the historical records show that England, Iceland, Greenland and western Europe enjoyed a warm period in mediaeval times. There are records of crops being grown in places in which by 1500 it had become too cold for farming. Climate scientists suggest that the change from the warmer mediaeval period to the cooler Little Ice Age is just the sort of slow climate change to be expected from the changes in the Sun's intensity, such as the changes in the sunspot cycle.

TEMPERATURES OVER THE PAST 2000 YEARS

The most easily accessible temperature indicators are historical records, essentially limited to the northern hemisphere, predominantly Europe and China. By their very nature they do not provide actual temperatures, but they certainly can reveal periods of warmer or cooler climate. These events are well documented in diaries, paintings and stories, such as the news report at the head of this chapter. They demonstrate climate change on the scale of several hundred years.

If you cannot get to a meeting and you want to cast a vote on some issue, you might appoint a proxy to vote for you. A climate scientist cannot find a fossilised mercury-in-glass thermometer to get a measurement of temperature for times long past, but there are other kinds of observations that can serve as proxies. In this case, a proxy is a measurement that indirectly estimates the annual average temperature. Provided whatever it is can be accurately dated and its changes compared to measured temperatures since 1880, it has the potential to be used as a proxy.

BOREHOLES

One of the more reliable temperature proxies comes from measuring the temperature down small boreholes into solid rock. The interior of the Earth is hot, but rocks are such good insulators that the Earth's surface temperature is set by the atmosphere. Down a borehole, the temperature slowly rises, at about 3°C per kilometre. If the surface temperature has been changing over a period of a few hundred years, that temperature rise is perturbed, and analysis can reveal the change in surface temperature. The Earth retains the memory of its annual surface temperature history, down to a depth of about 150 metres, and long-term average temperatures can be reconstructed with confidence as far back as 500 years and in some reconstructions for 20 000 years.

The method matches thermometer measurements in the recent past, so trust can be placed in borehole-derived temperatures for earlier times.[2] One analysis giving temperatures back 20 000 years has been published by Huang and colleagues at the University of Michigan in the United States.[3] They conclude that compared to the 1960–90 mean, current temperatures are about 1.5°C warmer than in the Little Ice Age, and more than half a degree warmer than during the Mediaeval Warming. Australian results indicate a cooling during the Little Ice Age of about half that experienced by the northern hemisphere.[4] Borehole temperatures are still the subject of careful analysis, and quite conceivably the details of the interpretation might change in the next 10 years, but the general magnitude of surface change would appear to be established.

TREE RINGS

For many years, botanists have recognised that the annual growth rates of trees, particularly those growing in cold climates, can be estimated from the thickness and density of their annual growth rings. Provided it has enough water, a tree's growth rate depends on the hours of sunshine it receives in the summer and on the temperature. Other things that can affect the climate are also recorded in tree rings, such as *El Niño* and *La Niña* events. By sampling trees from places where the main factor affecting growth is in temperature (mainly near-Arctic locations), most of the other things that affect tree growth can be separated out, allowing a connection to be made between tree-ring width and local summer temperature. This connection is quite clear during the first half of the 20th century, but there is divergence between tree-ring deduced temperatures and thermometer records after about 1960. According to D'Arrigo and colleagues: 'The principal difficulty is that the divergence disallows the direct calibration of tree growth indices with instrumental temperature data over recent decades.'[5]

One explanation for the divergence is that later 20th-century air pollution has dimmed the Sun enough to reduce tree growth, and so makes it look as though the temperature has fallen. Another is that moisture stress, possibly associated with increasing temperatures, has caused a change.[6] It was this issue that raised worldwide excitement in late 2009, when private email messages between scientists were illegally published, with the implication the scientists were leaving out results that did not conform to their hypothesis of global warming since 1960, namely tree-ring data that indicated global cooling.

A re-analysis of tree-ring data was undertaken by D'Arrigo, Wilson and Jacoby in 2006 (see Figure 9.1). Despite the divergence problem, they were able to detect the Little Ice Age and the Mediaeval Warming: for the latter, their estimated temperatures were of the order of 0.7°C cooler than in the late 20th century, although they advised caution in accepting these results until the divergence referred to above is better explained.[7]

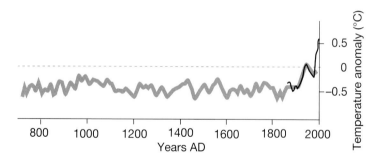

Figure 9.1 Generalised temperature trend from the year 700, based on tree-ring results. Baseline is scaled to the instrumental record 1960–90 (thin black line). Source: D'Arrigo et al., 2006.[7]

VARVES

In mountains of the near-Arctic, such as in Finland and Alaska, the spring thaw starts a rush of melt water down the valleys, which increases in the summer. The streams carry mud, silt and sand down the valley, leaving deposits on alluvial flats where the valley slope is gentler. In places, such a stream may enter a lake, and the coarser sediment, mainly silt, is deposited over the lake floor while some of the fine clay remains suspended in the lake water. As autumn returns the stream flow slackens and less and less sediment is brought into the lake, until by winter only a slow settling of clay is deposited on the lake floor.

Year by year, the layers of sediment accumulate: they are thicker, coarser and paler coloured at the bottom of each annual layer, and thinner, finer and darker at the top. The layers themselves are quite thin, perhaps only a few millimetres thick; they are known as 'varves' (Figure 9.2a).

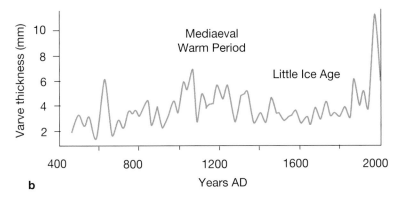

Figure 9.2 a) Varved shale from Montana, United States; one annual deposit is shown between finger and thumb. Source: Rod Benson. b) Generalised varve thickness trend from the year 500, based on the results of Loso et al., 2006.[8]

The thickness of an individual varve depends on that year's flow of water, which in turn depends on the summer temperature; a warmer summer leads to more rapid melting and so more sediment than the average, and vice versa for a cooler than normal summer. Measurement of varve thickness over a long period can give an indication of the climate change for that period.

One such study by Loso and colleagues[8] was made on lake sediment from Alaska (Figure 9.2b). They first established how the varve thickness changed with the annual temperature for the varve layers deposited in the past 100 years. They then were able to estimate the local temperature back to the year 400. Both the Mediaeval Warming and the Little Ice Age cooling were evident, as was a marked temperature increase in the 20th century. They concluded that the warmest estimated summer temperature of the Mediaeval Warm Period in this region was some 2°C cooler than the later 20th-century temperature estimate.

ICE CORES

In Chapter 6 we saw how ice cores can be used to estimate temperatures in times past. Using this method on tropical glaciers, Thompson and colleagues found a warming trend from 1700 to now, a cooler period from about 1400 to 1700, corresponding with the European Little Ice Age, and a slight warming centred at about 1100 (Figure 9.3).[9]

SEASHELLS

Another kind of proxy for temperature lies in the composition of the shells of sea creatures such as corals and free-floating plankton. From his analysis of sediments on the floor of the Bermuda Rise in the north

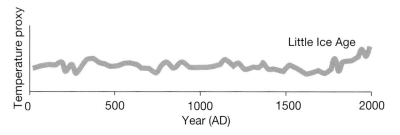

Figure 9.3 The temperature variation shown by tropical glacier ice-core proxies over the past 2000 years. Source: Thompson et al., 2006.[9]

Atlantic Ocean, Lloyd Keigwin from the Woods Hole Oceanographic Institution concluded that in this region the warming of the 20th century was not unprecedented.[10] Around 3000 years ago, the data indicate that sea-surface temperatures were at least a degree warmer than at present, and the Mediaeval Warm Period (~1000 years ago) was as warm or warmer than today.

Across the Atlantic, Paula Diz and colleagues from the University of Vigo, Spain, studied sediment from an estuary on the north coast of Spain.[11] They detected similar high sea-surface temperatures continuously from 1000 BCE to 1000 CE, but in a restricted region probably not representative of the open ocean. From 1000 CE the sea temperatures cooled until the end of the Little Ice Age (about 1750), followed by mild warming.

RECONSTRUCTION OF ALL PROXY TEMPERATURES

A more recent analysis by Mann and colleagues in 2009[12] concentrated on northern hemisphere data. Their results closely matched those from tropical glaciers and up until 1700 conformed to the longer-term trend of declining temperature related to the current Milanković cycle (Figure 9.4).

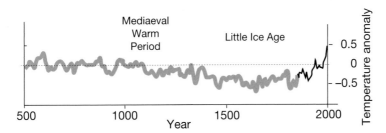

Figure 9.4 Northern hemisphere temperature reconstruction after Mann et al. (2009),[12] showing the average of all the proxy records as the green line, and the recent instrumental record in black compared to the 1850–1995 average (dashed line). The times of the Mediaeval Warm Period and the Little Ice Age are indicated.

On the basis of the historical records, the tree-ring measurements, varves, bore-hole temperatures and ice cores, several climate research groups have been able to reconstruct the global temperature changes over the past two millennia. Figure 9.5 (p. 148) is a simplified version of six such reconstructions, as published by Mann and colleagues in 2008.[13] The darker line broadly follows the reconstruction Mann and Jones published in 2003,[14] and the paler band covers the range of all reconstructions. They show the Little Ice Age and the Mediaeval Warming, and although the actual temperature interpretations vary a bit between the different groups, there is overall good agreement. It is particularly noticeable that the Mediaeval Warm Period, which in some earlier reconstructions was thought to have been as much as 2°C warmer than temperatures of the 20th century, is now thought to have been about much the same as the early 20th century and half a degree cooler than at the start of the 21st. Both Trouet et al.[15] and Mann et al.[13] attributed the Mediaeval Warm Period and the Little Ice Age to variations in one of the main oceanic drivers of climate, the North Atlantic Oscillation. A minimum in sunspot numbers around 1500 known as the Spörer Minimum, a second between 1650 and 1700 called the Maunder Minimum and the Dalton Minimum of 1800–20 probably contributed to the prolonged cooling of the Little Ice Age (see also Sunspot box in Chapter 3).

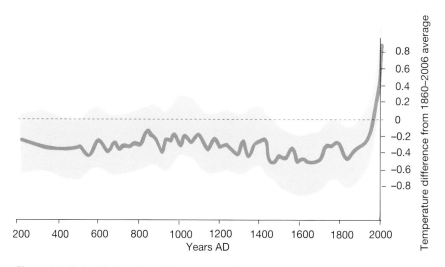

Figure 9.5 A simplified version of the summary of six global temperature reconstructions. The darker central curve follows the Mann et al. reconstruction of 2008,[13] while the lighter shading covers the range of all the other reconstructions.

The last five figures have all been graphs of the global temperature over the past 2000 years. Every one shows the same thing; a gentle, if rather irregular decline all the time until about 1800, then a sharp upturn, becoming steeper. These recent results confirm Michael Mann's 'Hockey Stick'.

Yet, some commentators suggest that this reconstruction of the past 2000 years is a fraud. Canadians Steve McIntyre and Ross McKitrick[16] were concerned about the statistical treatment of Mann's original data, and argued that a hockey stick graph was inevitable, and would result from that statistical method even using random data. This paper received several comments from the scientific community, of which one, while recognising the statistical problem, found McIntyre and McKitrick's concerns to be exaggerated,[17] while another found the statistical method did not have a significant impact.[18] In 2007, a multi-authored summary paper[19] assessed the recent temperature reconstructions, including the claims of McIntyre and McKitrick, and found that 'those claims were not well supported'. In the same year, a paper by Eugene Wahl and Caspar Ammann essentially laid to rest any doubts as to the robustness of the Hockey Stick's analysis of global temperatures.[20]

Since the last version of the Hockey Stick, two papers have added data that yield the same result. Von Gunten and colleagues used an analysis of sediment from a lake in Chile and concluded that the Mediaeval Warm Period was 1.2°C warmer than the Little Ice Age, and about the same as the 1940 global temperature.[21] A detailed examination of pollen extracted from Scandinavian lakes by Seppä and colleagues found essentially the same temperatures as the Hockey Stick.[22]

CARBON DIOXIDE

We saw in the previous chapter that CO_2 and global temperature are closely related. At the end of the Carboniferous period, and again at the end of the Cretaceous period, CO_2 reduction is thought to have contributed to global cooling. On the other hand, as the ice ages waxed and waned, parallel CO_2 variations followed and then amplified the changes caused by the Sun. Over the past 1000 years, CO_2 and temperature have continued to follow each other. Relationships between CO_2 and temperature from three Antarctic ice cores were published by a Swiss team led by David Frank in 2010.[23] From a high value of 285 ppm during the Mediaeval Warm Period, atmospheric CO_2 steadily declined until 1750, when it reached a low of about 275 ppm. Thereafter the trend has been sharply upward (see Figure 9.6, p. 150).

Results from analyses of ice cores can be compared with estimates of the atmosphere's CO_2 content made by measuring the CO_2-absorbing structures on fossil leaves. These structures, called 'stomata', become more abundant when the CO_2 levels drop and vice versa. Kouwenberg and colleagues measured such stomata in fossilised pine needles in lake sediment

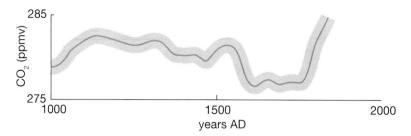

Figure 9.6 Atmospheric CO_2 content from 1000 to 1850, as determined from Antarctic ice cores. Source: Frank et al., 2010.[23] Note that the range in CO_2 is only about 10 ppm.

on Mt Rainier in the United States, dating back to 800.[24] Their estimates of CO_2 yield the same average CO_2 level of about 280 ppm, but in detail there is less agreement. The range in CO_2 is considerably greater in the pine needle study, reaching 320 ppm around 1400. These and similar studies tend to agree that from the end of the last ice age to around 1750, the atmospheric CO_2 content has fluctuated, probably by less than 25 ppm, around its average value of 280 to 290 parts ppm.

For 2000 years, the temperature has gently declined, following the slow change in the Sun's influence on climate, with fluctuations of no more than a few 10ths of a degree. In parallel, the amount of atmospheric CO_2 has fallen. We now have a very clear and robust long-term record of the Earth's natural climate variation. According to the Milanković cycles, we are now, or we should be, in a period of very slow cooling, with an ice age more than 50 000 years away.[25]

But we are not. We are in a period of unprecedented, rapid warming. Such warming is not the Earth's normal process of climate change as revealed by historical and geological study, and as is emphasised in a paper entitled 'Recent warming inconsistent with natural association between temperature and atmospheric circulation over the last 2000 years'.[26] Admittedly, in geological terms there is nothing special about today's global mean temperature. The Earth has experienced similar, warmer or cooler times over its vast history. What is special is not the temperature, nor even that it is changing. It is the speed of change that is special. Look back at the rate of warming as the last ice age ended. At its very fastest, the world warmed at a rate of 1°C every 1000 years. As far as palaeontology and geology can discover, temperature change as fast as 1.5°C a century has not happened in at least the past 2 million years; it has not happened over the time of *Homo sapiens*. Question 3 now has an answer: we have a fairly clear idea about how the climate has changed in the past. And now, knowing that, and that the climate is changing today, we can pose a fourth question: Is the current change normal in terms of climate history? The answer is NO. There is no known time when the whole Earth's climate changed at the pace it is changing today.

Those answers suggest the final question: what is the cause of current climate change? In the next chapter we will explore the final link in the logic that explains contemporary global warming and climate change.

SUMMARY

Topic	Observation	Key statistic	Conclusion
Temperature proxies	Estimate global temperatures over the past 2000 years.	Slow and steady temperature fall until ~1800.	Climate was following Milanković cycle expectation.
CO_2 levels	Slight decline for 1000 years, average 280 ppm.	Sharp increase from 1800.	Temperature and CO_2 were linked over this period, as they were over the ages past.
Hockey Stick	Slow temperature fall for 1800 years, sharp rise since.	Recent temperature rise is beyond all previous experience.	Today's climate change has no parallel in the past 2000 years.

FURTHER READING

Berger WH (2002) 'Climate history and the great geophysical experiment.' In G Wefer, WH Berger, K-E Behre & E Jansen (eds) *Climate Development and the History of the North Atlantic Realm*. Springer-Verlag.

Broecker WS & Kunzig R (2008) *Fixing Climate: What Past Climate Changes Reveal About the Current Threat – and How to Counter It*. Hill & Wang.

10

CARBON DIOXIDE AND METHANE

A smell of burning fills the startled air.

Hilaire Belloc

We have now reached a point in our enquiry where we have all the answers but one. We have seen that over the geological ages the climate has varied from hothouse to ice-house, and that the atmosphere has varied in parallel, from CO_2-rich to CO_2-poor. We have seen that small perturbations in the global temperature are magnified by several feedback mechanisms. We have seen that during the 20th century and into the 21st, both the temperature and the atmospheric CO_2 have increased, and we have seen the connection between the two. What we have not yet seen is the source of the CO_2.

WHERE DID THE CO_2 COME FROM?

The very accurate daily analyses of air from around the globe and from air trapped in ice have unequivocally shown us that CO_2 and methane have increased in our atmosphere over the past 200 years. We know how it is that these gases can affect the climate, so we need to find out where they come from and why they are increasing.

To start with CO_2, we can rule out the process that led to its pre-1800 level, namely warming of the oceans after the last glaciation. That glaciation ended about 10 000 years ago, and the global temperature has cooled slightly ever since. CO_2 levels rose over this period, from 260 to 280 ppm, thanks to delayed degassing from the ocean. That is a rise of one ppm in 500 years; the current rate of increase is almost two ppm in one year; that is, *one thousand times* as fast.

The graph in Figure 10.1 helps to put into perspective the sudden, rapid and continuing rise in atmospheric CO_2 since 1850. If the CO_2 is coming out of the ocean, something made it change its production rate

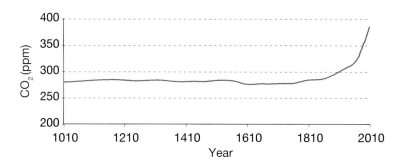

Figure 10.1 Atmospheric CO_2 levels compiled from Law Dome[1] ice cores and Mauna Loa records[2] for the past 1000 years.

very sharply. But there is a strong argument against the CO_2 increase coming from degassing the ocean.

Of the two more common isotopes of carbon, carbon from volcanoes is richest in ^{13}C, atmospheric and seawater carbon have a little less ^{13}C and plants have a lot less ^{13}C than the others. Over the past 30 years the ^{13}C proportion of atmospheric carbon has decreased by 10 per cent, and this provides an important clue to the source of the CO_2 increase.[3] Adding seawater CO_2 to the atmosphere will not change the proportion of ^{13}C to ^{12}C because ocean and atmosphere have about the same proportion. This means that whatever is the source of the rise in atmospheric CO_2 it must be a source poorer in ^{13}C relative to ^{12}C than the atmosphere, not one with the same proportion or with more. The decrease in ^{13}C in the atmosphere cannot have come from the ocean.

Perhaps the source is volcanic; volcanoes release a lot of CO_2, 300 million tonnes a year on average. The current annual increase in CO_2 in the atmosphere is about 10 billion tonnes a year. For volcanoes to be responsible they would have had to have been erupting at about 30 times their normal rate. Someone might have noticed. The largest land volcanic eruption since Keeling started his measurement of CO_2 in Hawaii was from Mt Pinatubo in the Philippines, in 1991. If Mt Pinatubo put vast quantities of CO_2 into the air all at once, the signal should have reached Mauna Loa in Hawaii, an island in the same hemisphere and ocean. In May 1992, Mauna Loa Observatory recorded 359.34 ppm CO_2. The levels fell steadily to 354.02 ppm in August that year. Pinatubo's main eruption in June left no trace whatsoever in the CO_2 record. Furthermore, the ^{13}C proportion of volcanic CO_2 is a little greater than that of the CO_2 in the oceans or atmosphere. Adding volcanic CO_2 should increase the atmosphere's $^{13}C:^{12}C$ ratio. As we saw, the reverse is true.

CO_2 is a well-known consequence of breathing, so conceivably it is the global increase in organisms that breathe out CO_2 that is responsible. Given the ever-rising world population, that might be the answer and it is easy to test. We breathe out about 1 kilogram of CO_2 each day. Six billion breathing humans over 365 days adds up to 2 billion tonnes, a large number but nowhere near the 30 billion tonnes of annual CO_2 addition. So I am afraid Jerry Adler's solution, that we all stop breathing for an hour,[4] would not be enough. In any case, the CO_2 breathed out by every creature on Earth was extracted from the atmosphere as grass (if we are carnivores) or fruits, nuts and cereals only a year or so earlier, so it is a zero-sum equation.

BURNING COAL, OIL AND GAS

When plants, coal or oil burn they convert carbon into CO_2. Burning coal and oil might be a source of CO_2 to the atmosphere. The timing is right for this hypothesis. Detectable increases in atmospheric CO_2 did begin more or less at the same time as the Industrial Revolution, a time when coal was first extensively mined and burnt. But is it possible that even the industrial effort of 6 billion humans could have any effect on the vast atmosphere that surrounds the Earth? Let us look at the numbers. In so doing, be careful to note that I have used billions of tonnes of CO_2. Most of the source data compiled by the Intergovernmental Panel on Climate Change are quoted in tonnes of *carbon*. The difference is a factor of 44/12 = 3.7, which takes account of the weight of the two oxygen atoms in CO_2.

The amount of fossil fuel mined or pumped, be it coal, peat, oil or gas, is well known to governments; they need to know because they tax it. It is easy therefore to determine how much of each fuel was used each year. A tonne of carbon in a fuel burns to make 3.7 tonnes of CO_2. The carbon content of each fuel is also a well-known figure, so a bit of arithmetic provides the necessary answer. In 2010 (the most recent year for which the figures are available), worldwide we burnt enough fuel to add almost 32 billion tonnes of CO_2 to the atmosphere.[5] Figure 10.2a shows the steady increase in CO_2 emissions from all fossil fuels.

The next graph (Figure 10.2b) shows atmospheric CO_2 measurements since 1890, and Figure 10.2c shows the 10-year averaged global temperature anomaly. Three graphs, one story. Atmospheric CO_2 and temperature have risen broadly in parallel with the CO_2 emissions from burning fossil fuels, although the temperature rise has been modified by other factors as explained in Chapter 2.

Let's return to the changes in the light and heavy kinds of carbon atoms. When plants take up CO_2 they strongly prefer ^{12}C over ^{13}C, with the result that coal and oil have a very low $^{13}C:^{12}C$ proportion. Burning such 'light' carbon (light because it has less of the heavier form of carbon than the atmosphere) will add less ^{13}C to the atmosphere than ^{12}C, therefore the atmosphere's content of ^{13}C should fall as fossil fuel CO_2 is added, exactly as is found. Thus, there are at least two lines of evidence that suggest the increase in atmospheric CO_2 might be linked to fossil-fuel burning: both the production of CO_2 from fuels and the atmospheric content of CO_2 have risen in parallel for many years, and the composition of the atmosphere's carbon is changing in a way that is consistent with the addition of fossil carbon.

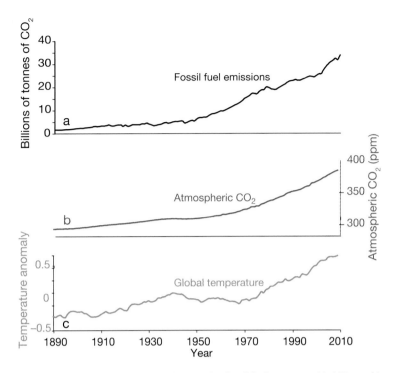

Figure 10.2 a) Annual total CO_2 produced by burning fossil fuels, measured in billions of tonnes of CO_2.[6] b) Atmospheric CO_2 content in ppm. Source: as for Figure 10.1. c) 10-year averaged global temperature anomaly. Source: HadCRUT3.

And there is a third line of evidence. Most of the world's production of CO_2 comes from the major industrial regions, all in the northern hemisphere. Following the creation of extra radioactive carbon during the era of surface atom-bomb tests, scientists were able to find out how long it takes for the circulation of the atmosphere to move northern-hemisphere CO_2 to the southern hemisphere; it takes one or two years, depending on the latitude of the source.[7] Comparing the amount of CO_2 in the atmosphere over Alaska and over the South Pole shows that it takes about two years or so for the South Pole to catch up with Alaska (Figure 10.3, p. 158). Clearly, whatever is causing the atmospheric CO_2 to rise, it is not occurring uniformly around the globe; it is largely happening in the northern hemisphere.

When fuels are burnt, oxygen is consumed. Since atmospheric oxygen stands at 21 per cent and CO_2 at 0.04 per cent, the depletion in oxygen is not going to make us short of breath, but it is measurable. Since 1992, Ralph Keeling and colleagues at the Scripps Institution of Oceanography in La Jolla, California, have been measuring the air's oxygen, and it is falling in step with the rise in CO_2.[8]

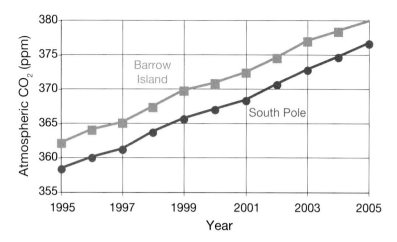

Figure 10.3 Atmospheric CO_2 measurements form Barrow Island, Alaska, and the South Pole. South Pole values reach Barrow Island levels about two years later. Source: Carbon Dioxide Information Analysis Center.

Not all the CO_2 produced since the Industrial Revolution has been from burning fossil fuels. Some is released into the atmosphere by cement manufacturing. Cement is made by grinding up limestone and clay, then heating the mixture until a chemical reaction liberates CO_2 from the limestone and produces cement. The amount of CO_2 put into the atmosphere from cement production in 2010 is estimated at just over 1.5 billion tonnes.

There is also significant release of CO_2 when land is cleared for agriculture, primarily through burning the original vegetation, but also from drying the soil and oxidising some of the soil carbon. According to an analysis by a group of US scientists, each hectare of cleared land in temperate climates released 60 tonnes of carbon to yield about 4 tonnes of dry crop. In the tropics, the figures are 120 tonnes of carbon released for about 2 tonnes of crop.[9] Through clearing and burning, in 2010 we added another 3.5 billion tonnes of CO_2.[10]

Adding the 32 billion tonnes of CO_2 from burning fuel, 1.5 billion tonnes more from cement manufacturing and another 3.5 billion tonnes from land clearing gives an annual input of human-derived CO_2 of 37 billion tonnes. How does that amount compare with the annual increase in atmospheric CO_2, which in 2010 stood at 2.4 ppm (by volume) of the atmosphere? That apparently small amount multiplies to 18 billion tonnes.[11] But 18 billion tonnes is only half of the CO_2 produced by burning fossil fuels and changing land use. Where is all the rest? To explain that, we first need to look at what is known as the Earth's carbon cycle. In Chapter 4 I described the long-term fate of CO_2 as it moves out of volcanoes into the atmosphere, then the oceans, then into shells and finally into rock. There is also a short-term cycle that is more significant to our human life span, and one that we appear to be able to change.

THE CARBON CYCLE

It is possible to identify places where carbon is stored. Not counting rocks – mainly limestones – the ocean is the biggest store of carbon, containing 91.5 per cent of all carbon that moves between the ocean, soils, land and air. Soils come second with almost 4 per cent, then come plants and ocean creatures, making up 3 per cent, and the atmosphere comes last with a bit less than 2 per cent. Soil, perhaps surprisingly, holds twice as much carbon as the air does (see Figure 10.4). Soil carbon is in the form of roots and organisms as well as patches of limestone that have been formed by rock weathering.

The way the carbon moves about is called the 'global carbon cycle'. In pre-industrial times there was a balance as CO_2 moved between land, sea and sky annually. That balance has now been disturbed: of the 700

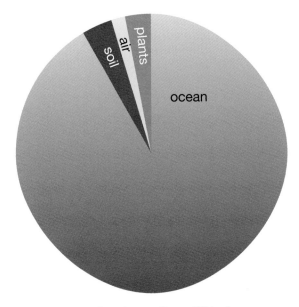

Figure 10.4 Earth's short-term carbon storages. Source: Wikipedia.

gigatonnes of carbon in the atmosphere (as CO_2), just over 90 gigatonnes move into the oceans and a bit less moves back out of the oceans each year (Figure 10.5). The carbon that goes into the ocean is used by organisms, while the carbon that comes out of the ocean by degassing (that is equilibrium exchange, not the result of warming the ocean) almost replaces the CO_2 that went in. That means that overall the ocean is presently a sink for CO_2, not a source.

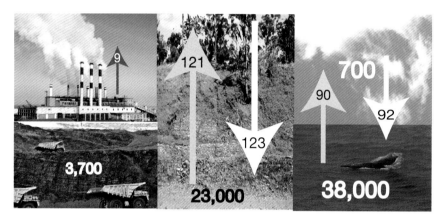

Figure 10.5 Carbon stores (larger numbers and larger print) in fossil fuels, the land, the ocean and the atmosphere, and carbon exchanges between them (smaller print numbers on arrows). All numbers are billions of tonnes of carbon, and all are approximate. Source: data from IPCC Fourth Assessment Report, 2007.

Similarly, carbon is cycled between the land and the atmosphere by plants, animals and soil processes. Like the ocean, the land is a sink for CO_2. However each year the ability of the land to take up CO_2 changes depending on the weather; wet years encourage plant growth and more CO_2 is taken up, warmer, wetter years accelerate rock weathering and this also takes up CO_2. The *La Niña* year of 2008 was a good year for the land; it is estimated that about 35 per cent of all the CO_2 we added to the atmosphere that year ended up in plants and the soil.[12] A further 25 per cent entered the ocean, and the remaining 40 per cent of the CO_2 stayed right where we put it: in the air (Figure 10.6). But two years later the numbers were different. For 2010, the land only took up 25 per cent of the added CO_2, so that half of what we produced by burning fuels remained in the atmosphere.[5]

Over the past 50 years the ocean has been taking up proportionately a little less of the CO_2 than we emitted. This suggests that we are adding CO_2 to the atmosphere faster than the ocean can absorb it, whereas,

Figure 10.6 What happens to CO_2 emissions. In 2008, 60 per cent was taken up by plants, soil and the ocean, while the rest remained in the atmosphere. Source: percentages from Le Quéré et al., 2009.

though there is considerable fluctuation, on average the land has not changed in the proportion it takes out of the air.

There are not many places in the carbon cycle where we can have an influence. Not burning fossil fuels would be one way to change the balance. Increasing the amount of CO_2 plants take up, and making sure the plant carbon does not burn, either quickly in a fire or slowly by decomposition, is a second carbon sink we can affect. The deep-ocean storage is not something we know how to modify, though it has been suggested that adding iron to the ocean would 'fertilise' the growth of plankton and so increase the ocean's uptake of CO_2.[13] A dissenting paper by four scientists from the United States published in *Nature* in 2009 has a different view: 'Adding iron to the ocean is not an effective way to fight climate change, and we don't need further research to establish that.'[14]

METHANE SOURCES

After water vapour and CO_2, the next most important greenhouse gas is methane, which has the chemical formula CH_4. Methane in the atmosphere has more than doubled since pre-industrial times and for it, too, we need to find the source of the increase.

One source of methane is marshes, or wetlands. Marsh gas is methane, and willow o' the wisp is (probably) naturally ignited methane burning over a marsh. Rotting vegetation, in general, releases methane, as do a variety of other biological activities that occur in the absence of oxygen, such as in the gut of termites and cows, particularly, and also in other mammals, including some humans. All of the methane that is not directly attributable to human activity; that is, wetlands, termites and wild animals mainly, amounts to about 40 per cent of the annual methane release. The remainder results from human activity in various forms.

Rice agriculture contributes about 10 per cent of the total methane released. The gas that kills coal miners – by asphyxiation or explosion – is methane. Coal mining releases a lot of methane into the atmosphere, as do the gas and oil industries, together amounting to about 20 per cent of the annual methane release. Landfill and biomass burning appear to account for the remainder.

The exact, or even fairly close, values of all the sources and sinks for methane are not well known, but it is clear there has been a steady rise in atmospheric methane from 700 parts per billion (ppb) before the Industrial Revolution to its current level of 1750 ppb. From 2000 to 2006, the atmospheric methane levels remained almost constant, and then, with no adequate explanation yet being proposed, started to rise sharply again (see Figure 10.7a, which is the Antarctic Mawson base record from 1985).

Methane sources (rounded to the nearest 5 per cent):[15]

Wetlands	30%
Coal, oil and gas extraction	15%
Animals, mostly domestic	15%
Rice cultivation	10%
Land clearing and forest fires	10%
Landfill and waste	10%
Termites	5%
Volcanoes, the ocean	5%

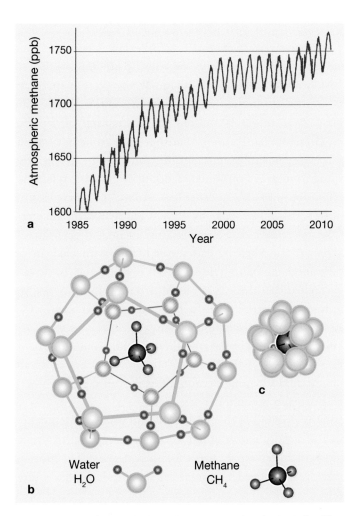

Figure 10.7 a) The trend of atmospheric methane measured at the Australian Mawson Base in Antarctica. Source: CSIRO Marine and Atmospheric Research. b) In this model of a methane ice crystal, the water molecules are the blue balls (oxygen) with smaller brown balls (hydrogen) attached; the methane is the grey (carbon) and brown (hydrogen) structure within the 'cage'. In this figure the sizes of the atoms comprising the molecules are much reduced to show the structure of the cage more clearly. c) The same model as in b), drawn much smaller but with the atom sizes in true proportion, showing that the methane molecule is too big to escape from the ice cage (unless it melts).

Climate scientists are concerned and some are alarmed about methane that is trapped in sediment below the Arctic and Antarctic oceans in a form known as methane ice (methane clathrate, also called methane hydrate), where the methane molecule is trapped in a frozen cage of water molecules in a structure shown in Figure 10.7b. A five-year study of methane release in the East Siberian Arctic sea[16] found this area alone was releasing as much methane as previous estimates for the entire world's release of ocean methane. While there is speculation that the renewed increase in the rate of methane addition to the atmosphere found since 2007 might be related to the melting of these methane clathrates, climatologists caution that there is insufficient knowledge of even the recent past emissions from this source. One group of scientists that is trying to improve our understanding of methane release is headed by Igor Semiletov of the International Arctic Research Centre at the University of Alaska Fairbanks in the United States. In December 2011 he told *The Independent* that over the 20 years he had been surveying the East Siberian Arctic Shelf he had 'never before witnessed the scale and force of the methane being released from beneath the Arctic seabed'.[17]

It is estimated that there is as much carbon in methane ice under the ocean as in all the Earth's fossil-fuel deposits.[18] As we saw in Chapter 8, geologists suspect rapid temperature rises such as that which occurred 55 million years ago might have resulted from the rapid release of methane from such ice crystals. Theoretically, the thawing of methane ice could trigger rapid warming because of its powerful greenhouse effect. It happened once and it could happen again.

Methane is oxidised in the air to CO_2 and water quite quickly; its residence in the atmosphere is about 10 years (compared to the residence time of CO_2 of anywhere between 5 and 100 years). Therefor, its greenhouse effect is short lived, although once oxidised it adds a little to the CO_2 levels.

Since the start of the Industrial Revolution, approximately 1.25 trillion tonnes of CO_2 have been added to the atmosphere. Of that, a little less than half is still in the atmosphere. A lesson we have learned from the ice ages is that, with a little push from the Sun, CO_2 rose from 180 ppm to 280 ppm while the global temperature rose by 6°C. That was enough to melt ice sheets over Europe and North America that were 3 kilometres thick. Our ancestors saw amazing changes as the ice slowly melted. Over the 20th century, with a big push from us, CO_2 has risen from 280 to 380 ppm. Now it is our turn to see what global warming can do.

SUMMARY

Topic	Observation	Key statistic	Conclusion
CO_2 sources	Not the ocean, not volcanoes	CO_2 rise matches fuel consumption	Burning fossil fuels is the source of rising atmospheric CO_2.
The carbon cycle	CO_2 is taken up by plants, the soil and the ocean.	60% of our fuel emissions are absorbed, 40% remain in the atmosphere.	Earth's ability to absorb CO_2 is declining.
Methane	40% emitted by animals and wetlands, 60% from coal and oil exploitation, huge store in Arctic waters as methane ice.	30 times more potent as a greenhouse gas than CO_2.	Time bomb.

We now have an answer to the final question: What *is* the cause of current climate change? Answer: The climate is changing and the reason, the sole reason, is the quantity of greenhouse gases added to the atmosphere since the Industrial Revolution. Greenhouse gases have blocked enough heat to cause a rise in global temperature, and that temperature rise creates change in oceanic and atmospheric behaviour: climate change.

We have already seen some consequences of global warming. What might we expect in the future?

We know that climate change came about through the burning of fossil fuels. Is there anything we can do to reduce or remove the excess CO_2?

It is time to look to the future, but there is a road block. There are people, influential and vociferous, who are convinced that all these conclusions are wrong, who deny that climate change is happening, or that it is of any consequence. In the next chapter, I address their concerns, their evidence and their arguments.

FURTHER READING

Berner RA (2004) *The Phanerozoic Carbon cycle: CO_2 and O_2.* Oxford University Press.

  A SHORT INTRODUCTION TO CLIMATE CHANGE

11
DENIAL

'Contrariwise', continued Tweedledee, 'if it was so, it might be; and if it were so, it would be: but as it isn't, it ain't. That's logic.'

Lewis Carroll

THE SCIENCE OF NO CLIMATE CHANGE

At the start of this book I wrote: *I will not be selecting the reports that support (or deny) any particular view of the subject, but I will be selecting reports that have the authenticity that comes from peer review and then public exposure.* Now we are approaching the end of the book, and you must be asking: 'Where is the science that contradicts the theory of climate change?' You may well have read contrary conclusions to those presented here, in the newspapers or on the internet or in other books. Surely there is a body of science that underpins those views? So why is none of that included in this book?

The answer is because there is no such body of knowledge. I looked. I searched extensively. After finding very little of repute, I turned to the writings of those who declare themselves to be sceptics on the subject.

Two scientists who have become prominent because they deny that global warming is caused by the burning of fossil fuels are the retired Australian professors of geology, Ian Plimer and Bob Carter. They approach climate science from the perspective of geologists who understand how the Earth's natural processes have led to climate change in the past, and are of the view that if there is any current change in climate we must look to natural causes. This is a sensible, scientific attitude; indeed, one that is espoused by all climate scientists. The natural causes of climate change are always and ever included in their research.

The main target of those who deny climate change caused by the burning of fossil fuels is the Intergovernmental Panel on Climate Change (IPCC). The sceptics regard this body as one that is politicised, as one that has made up its mind about climate change and that it is now doing everything it can to 'sell' the story. According to Carter: 'The declared intention of the IPCC was to provide disinterested summaries of the state of climate science as judged from the published, refereed scientific literature.' This is correct. But Carter continued: 'Accordingly in four successive assessment reports in 1990, 1996, 2001 and 2007 the IPCC has tried to imprint its belief in dangerous human-caused warming on politicians and the public alike, steamrollering relentlessly over the more balanced, non-alarmist views held by thousands of other qualified scientists.'[1] (page 28) This was Carter's own opinion, not one that is supported by evidence. Almost none of those 'thousands of other qualified scientists' were qualified in climate science, and they have so far failed to provide any conclusive evidence one way or the other. A handful of climate scientists have published papers arguing against human involvement in climate change, and I will comment on some of these later in this chapter and the next.

For all their criticisms of the IPCC, the writings of Plimer and Carter make it appear as though they have not actually read the reports. I suspect that they have, but what is to be made of this sentence by Plimer: 'The whole basis for human induced global warming can be found in just one chapter (Chapter 9) of the IPCC Report AR4.'[2] This assertion by Plimer is plain wrong. In the IPCC report, four chapters detailing the observations of climate change are followed by a chapter on palaeoclimates, then three chapters explaining climate change, including 'human-induced global warming'. Carter writes: 'The IPCC concentrates its analysis of climate change on only the last few hundred years, and has repeatedly failed to give proper weight to the geological context of the short, 150-year long instrumental record. When viewed in geological context, and assessed against factual data, the greenhouse hypothesis fails'[1] (page 29). This, too, is incorrect. All of Chapter 6 and much of Chapter 9 of the IPCC 4th Assessment Report are devoted to the geological context of climate change.

Carter and Plimer emphasise the point that science does not progress by consensus and they declare that accepting the consensus is unscientific. It is true that science does not advance on the basis of consensus, but most of the great leaps of science, though counter to the then-accepted wisdom, were built on the knowledge embraced by that wisdom. Science can only go backwards when it ignores the documented, challenged and eventually accepted wisdom of many, many scientists. Similar results emerging across a wide range of research topics, and a broad consensus about climate change reflected in the IPCC reports, has not stopped scientists from critically evaluating each others' work. Any scientific consensus must be flexible, ever-changing and able to stand up to scrutiny. Assertions are not scrutiny.

A very quick review of 20th-century history of climate science might help to understand how it is that the IPCC reports have come to represent the collective view of most climate scientists. Until about the mid-20th century, climate scientists were not really researching climate change; rather, they were investigating climate as a phenomenon: what affects it, the history of climate, why the ice ages come and go, and so on. The greenhouse effect has been known since the 19th century (see Chapter 4). In 1937, Guy Callendar, a British engineer, wrote of the artificial production of CO_2 and its influence on temperature.[3] Callendar reported a global temperature rise in the period 1900 to 1930 amounting to some two-10ths of a degree, and attributed this rise to the addition of CO_2 to the atmosphere from the burning of coal, oil and peat. In the mid-1950s, Keeling's analyses of CO_2 confirmed Callendar's earlier suspicions, and this was followed by the work of such people as Wally Broeker, Michael Mann, James Hansen, Tom Wigley and Phil Jones. Collectively, these men shook the foundations of 'steady as she goes' climate science with the outrageous claim that we, the human race, had the power to change the climate. You can be sure that the first papers published on this subject were not instantly acclaimed by all climate scientists. Nor were these scientists flooded with grant money. There were challenges, there were questions; the studies were duplicated and triplicated. New tests of the ideas were thought up and applied. Gradually, over time, the scientific community began to see the reality in those early results and interpretations. Today, the majority of climate scientists know that the data have been checked, that the interpretations have been made properly and logically and that the conclusion that the Earth is warming due to the addition of CO_2 to the atmosphere is fully justified.

DENIALISTS

However, there continue to be those who refuse to accept the evidence, and here I address the views of three of these authors since, if anyone knows of scientific research that cogently argues against climate change and global warming, they do. Here are examples of what I found.

The first comes from Professor Ian Plimer's 2009 book *heaven+earth*.[4] I chose this work because we need to develop an economy with low carbon emissions, and we will not do that if many of us simply do not believe there is any reason to do so. Plimer's book makes for very comfortable reading. It tells us not to worry, that burning coal and oil does no harm to the planet, does not change the climate and has not increased the temperature. If convinced by Plimer, readers will not support policies to reduce carbon emissions, industry will carry on with 'business as usual' and, in the words of James Hansen of NASA's Goddard Institute for Space Studies, the future of our grandchildren will be stormy, indeed.

The first figure in *heaven+earth* is the one I reproduced as Figure 1 in Chapter 1. The subtitle to Plimer's book is: *Global Warming: The Missing Science*. However, there is missing science in his graph. For example, Plimer could have included temperature data from 1975 and he might have included real data for 2008 as he did for his Figure 4. The graph in Plimer's Figure 1 gives the impression that he used the 2008 datum, but he has not, because the point for 2008 (shown as a star on Figure 1.1, see page 3) falls well above his downward trending line.

Further on Plimer stated that: 'Global warming and a high CO_2 content bring prosperity and lengthen your life.' This is a curious statement for Plimer to make. Remember, he is denying that global warming exists (his Figure 1). CO_2 levels have not been higher than about 280 ppm for at least a million years, so how can he know they bring prosperity? How can he tell they lengthen your life?

Some of the evidence Plimer presents to deny a connection between CO_2 increase and temperature change is simply false. In his Figure 4 he draws the lower troposphere (lower atmosphere) temperature measured by the MSU satellite (as well as the HadCrut3 temperature data). It also has a line purporting to show CO_2 increase over the period shown (2002–9). This graph is an almost exact copy of one published with inadequate referencing on several websites, including one by the US 'Centrist Party'. The difference between them is that the dates on the website charts show 1998 to 2008, while Plimer's dates show 2002 to 2008 on the same chart. This makes Plimer's CO_2 slope much steeper than in reality, and completely negates the temperature data. I cannot find the MSU lower troposphere data set that exactly matches the curve of Plimer's figure; the closest are for the global data (latitudes minus 70°S to 82.5°N). Here, in Figure 11.1, I show the MSU data as the zigzag line and the Mauna Loa CO_2 annual average trend line. The correct dates for this figure are shown above Plimer's dates (in italics). He argues that the CO_2 trend does not match the flat MSU temperature trend. But when you look at the longer period (Figure 2.8, see page 23), there is a significant temperature increase, and it more or less matches the Mauna Loa record of CO_2 increase, as we have seen in earlier chapters.

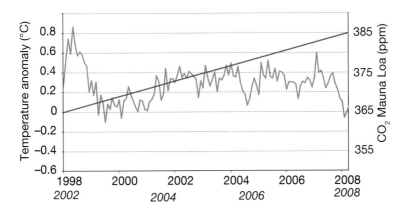

Figure 11.1 Reconstruction of Plimer's Figure 4. The zigzag line is the MSU-TLT lower troposphere data for the globe (NOAA, US National Climatic Data Center). The straight line is the averaged Mauna Loa CO_2 trend. On the horizontal axis, the dates on the upper row are correct, while dates on the lower row (in italics) are as Plimer presented his figure.

There are many criticisms of Plimer's book available on the internet. One you might read was written by scientist Ian Enting from The University of Melbourne (www.complex.org.au/tiki-download_file.php?fileId=91).

In 2011, Plimer updated *heaven+earth* with a book designed to appeal to 'punters, pupils and parents'. Much of the misinformation from his first climate book is repeated in the second, and he concludes with 'one hundred and one questions for your teachers'. These questions reveal much about the denial of science. One technique is to include a fallacy in the question in such a way as to make it appear to be fact. Here are a few examples.

> 13) Why is there no correlation between global warming and carbon dioxide yet there is a correlation between solar activity and temperature?

Clever question, in the same vein as the apocryphal lawyer's question to the defendant: 'Have you stopped beating your wife yet?' The first part of the question assumes a falsehood; in fact, there *is* a correlation between global temperature and atmospheric CO_2 going back as far as you like. The correlation is particularly close – or particularly well documented, anyway – during the past million years (see Chapter 8).

> 24) If carbon dioxide drives global warming, how is it that we have had six major ice ages in the past, yet atmospheric carbon dioxide was far higher then than now?

Plimer says: 'Don't think you will get an answer to this question.' Well, nor should you, since it is another with a false premise. Atmospheric CO_2 was never 'far higher than now' during the ice ages (see Chapter 8). You have to go back 24 million years to find a recorded CO_2 level above 300 ppm.[5]

> 30) Can you please show me how 3% of annual emissions of carbon dioxide, that is the human emissions, drive climate change and the other 97% do not?

Plimer's answer is that 'this has never been shown, … and the whole story of human induced global warming is just nonsense'. The nonsense, of course, is in the question. Human emissions only add about 0.7 per cent to the atmosphere's CO_2 content. All the CO_2 in the atmosphere contributes to the establishment of the basic global temperature level, keeping the planet warmer than the ice-box it would otherwise be. In the same vein as this question, Plimer and others also reassure their readers that the air no longer contains many of the particular molecules of CO_2 that came out of chimneys. According to Plimer, this means that fuel-derived CO_2 cannot be causing global warming because those molecules no longer exist. This is just ridiculous. The CO_2 cycle is not picky, and chimney smoke CO_2 is cycled as readily as all the rest. The molecules absorb heat regardless of their origin.

48) Why could the Northwest Passage be navigated in the 1930s and 1940s in wooden boats, yet it could not be navigated in the late 20th century warming?

49) In 1903 Amundsen passed through Canada's Northwest Passage from the Atlantic Ocean to the Pacific. If the planet is warming, why is this not possible now?

50) I heard that 2010 was the hottest year since records have been kept. The Northwest Passage is closed by ice, yet it was open in the 1930s. Was 2010 really the hottest year on record?

In these three questions, Plimer appears to bolster his case that 20th-century global warming is no different from earlier periods of warmth. Sailing through Canada's Northwest Passage has been the dream of Arctic explorers since the mid-18th century. The first successful passage was by Norwegian explorer Roald Amundsen in 1903; it took him three years. In 1937, E.J. Gall from the Hudson Bay Company was able to transit one section of the Northwest Passage – from Cambridge Bay through Bellot Strait to Port Kennedy. Three years later, another experienced and determined Arctic sailor, the Norwegian-born Canadian Henry Larsen, achieved a complete crossing, but he had to overwinter halfway through the passage. In 1944 Larsen travelled the entire passage, the first explorer to do so in a single season. As you may recall from Chapter 2, 1937–43 was the warmest period of the century to that point. Since the year 2000, cruise ships, yachts and other small vessels, and even an inflatable vessel, have managed the trip through the passage. In 2010, the Northwest Passage was certainly not closed by ice, because at least 18 vessels traversed it. In 2011, cruise companies were advertising tourist trips through the passage, so it clearly is still possible (in summer).

One of Plimer's answers reveals why he so denies most of modern climate science.

27) Can humans change climate?

Plimer states that: 'The natural forces are far greater than anything humans can ever muster. The only answer is no.' Here, I think, is the crux of Plimer's polemic. He simply cannot accept that one of the processes of natural climate variation has been enhanced by the burning of fossil fuels and that this addition has increased the ever-present absorption of outgoing heat by greenhouse gases. The same mindset is evident in *heaven+earth*, where he writes: 'If we humans, in a fit of ego, think we can change these normal planetary processes, then we need stronger medication.' Humans have not added a new process; over the past 150 years we have augmented, by about 30 per cent, one that was always there.

Professor Carter from James Cook University in Townsville, Queensland, wrote *Climate: The Counter Consensus*, which was published in 2010. One theme permeates this book: the 'wickedness' of climate scientists, as evidenced, for example, by the so-called 'Climategate affair', now completely refuted as I mentioned in Chapter 1. This concept, essentially a battle between good and evil, is subtly reinforced again and again by Carter's division of the world of climate science into two groups: the evil IPCC and the righteous 'independent' scientists. Plimer used the same technique to sway emotion. However, the IPCC is not an organisation that *does* science; rather, it is an organisation that collates and summarises scientific results on climate.

Carter attacks climate science most vigorously on two fronts: on the subject of temperature, he doubts the reliability of instrumental records and argues for cooling since 1998; on CO_2, he doubts that the rise is due to the burning of fossil fuels. Carter supports his arguments with a number of references to publications, and these merit our consideration.

The instrumental record of temperature since 1850 is one part of climate science that is vigorously challenged on the basis that the urban heat-island effect and faulty statistical treatments render much of it invalid. 'To attempt to assess the dangers of climate change on the basis of arbitrary lines through temperature series that represent 1, 2, 5 or 11 climate data points, as the IPCC largely does, is clearly a futile exercise.' (Carter has arbitrarily decided that it requires 30 years to characterise a 'climate', therefore temperature measurements over 150 years only contain five data points, or five lots of 30 years.) This futile exercise is not so futile when he wants it not to be. Using the same data he alleges were 'corrected', 'manipulated' and 'tampered with' for the Arctic region, Carter claims that the temperature, which he accepts rose from 1960 to 1990, is not on a generally upward trajectory, but rather is following a rising and falling, wave-like pattern, with cooling predicted until 2030.

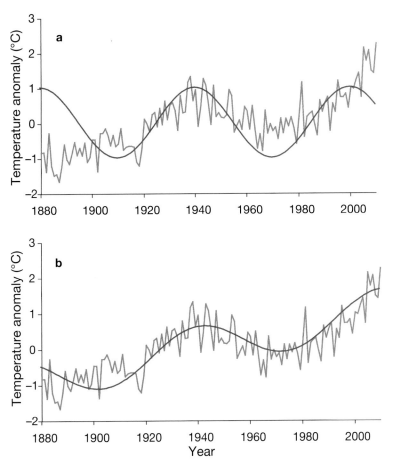

Figure 11.2 a) Land temperatures north of 64°, from the Goddard Institute for Space Science (zigzag line) with Carter's 60-year periodic variation fitted (by me) to the 1940 maximum (smooth line). As you can see, 'it fits where it touches'. b) The same temperature curve with my own (unscientific) black line made by adding an annual temperature increase of 1.5°C per century to a 70-year periodic variation of amplitude 0.6°C.

He bases this on only three data points (his concept), from 1900 to 2006. Then he draws his own line, presumably futilely, through the 'tampered with' data to conclude that we are now headed for 30 years of cooling. I hope he is right. I am unable to reproduce Carter's Figure 25 because he gives no source for its temperature data, but Figure 11.2a reproduces the temperature data for the Arctic region (observations from the Goddard Institute for Space Studies north of 64°N latitude), including the timespan presented by Carter, but adding the 20 years before 1900.[6] With those 20 years and the benefit of an extra 4 years (2007–10) of global temperature data, the suggestion of a simple 60-year cycle (the black curve) looks rather less probable; Carter could have made these extensions to the published work but obviously chose not to. I can achieve a much better-looking fit by adding the average temperature increase over the period 1880 to 2010 to a less pronounced 70-year cycle, but there is no good evidence for this being real (see, for example, the section on the Atlantic Multidecadal Oscillation in Chapter 3). I include it in Figure 11.2b only to demonstrate how easy it is to fit your own pet theory to limited data.

Carter is correct in recognising small rises and falls in temperature over periods of several decades; all the reconstructions of global or regional temperature have them. One source may be the Atlantic Multidecadal Oscillation I described in Chapter 3. The low point around 1970 was explained in Chapter 3 as a consequence of sulphate aerosols reflecting sunlight and cooling the Earth, and was probably also affected by the Atlantic Multidecadal Oscillation (or the former affected the latter). Most of the atmospheric pollution causing cooling was generated in the northern hemisphere, and northern hemisphere temperatures show the cooling most strongly. The Australian averages (Figure 2.3, page 18) show almost no such effect, and in the global averages (Figure 2.4, page 19) cooling was less pronounced than the northern Arctic data show.

To emphasise the cooling that Carter believes has occurred since 1998, he refers to a paper published in 2006 titled 'Recent cooling of the upper ocean'. He includes this paper in his reference list at #118, as well as a following paper by the same authors, titled 'Correction to the recent cooling of the upper ocean'.[7] In the second paper the authors report bias in the instrumental data used in the first paper and conclude: 'Both biases appear to have contributed equally to the spurious cooling.' Spurious cooling, but not so acknowledged by Carter. This cooling has, according to Carter, led to sea-level fall since 2004. But not according to the CSIRO, whose data from tide gauges and satellite measurements show steadily rising sea levels from 2004 to 2010 (Figure 7.4, page 112).

To further promote the case for late 20th-century cooling, Carter writes: '... the Arctic region was warmer in the early 1940s than in the 1990s, and has in fact actually cooled since 1920'. This statement was sourced from a paper by Polyakov and colleagues,[8] which concludes: 'Arctic temperatures in the 1930s – 40s were exceptionally high so that from the 1920s forward (up to 2000) the data have a small but statistically significant cooling tendency.' The authors further conclude: 'Since 1875 Arctic air temperature shows warming with an average rate of 0.09°C per decade' and '… the rate of Arctic surface air temperature increase in 1875–2001 is twofold compared with the Northern Hemisphere trend'. There has been considerably more work done on Arctic surface temperatures than this, summarised by Rajmund Pryzbylak from the Nicolaus Copernicus University in Poland.[9] Pryzbylak depicts temperature trends over each of six Arctic regions and averages over the four seasons; all seasons and all but one region (the Baffin Bay region has a flat overall trend) show warming from 1950 to 2000, and he notes: '… the period from 1995–2005 was the warmest since at least the 17th century'. Pryzbylak concludes that temperature variations for the period up until about 1970 were largely driven by natural factors, but suggests three reasons for the marked later 20th-century warming: natural causes, such as decreased solar reflection as snow and ice melted and incursion of warm Atlantic waters, greenhouse gas loading, or a combination of these factors.

Turning to the evidence of global temperatures before the introduction of reliable thermometers, Carter attempts to play down the significance of the 'Hockey Stick' (see Chapter 9) with a short but incorrect comment that the Mann, Bradley and Hughes papers 'were based on a statistical analysis of about 183 tree ring records across the Northern Hemisphere' (page 152). Here is what Mann, Bradley and Hughes actually wrote in their seminal 1998 paper: 'We use a multiproxy network consisting of … annual resolution dendroclimatic (meaning largely tree-ring data), ice core, ice melt … combined with other coral, ice core, dendroclimatic, and long instrumental records.'[10] In 2003, Mann and Jones extended the analysis with the help of newly published temperature reconstructions, of which tree-ring analyses comprised a little more than half.[11] In both the 1998 and 2003 work, the distribution of the records was certainly dominated by the northern hemisphere, but southern hemisphere records were included.

Another argument offered by Carter in his efforts to demonstrate the falsity of 'consensus' climate science involves denial of the proposition that atmospheric CO_2 gradually rose during the 20th century from a pre-industrial figure of 280 ppm to the unprecedented (in at least the past 24 million years) level of 380 ppm. Many of Carter's arguments about CO_2 appear to be drawn from one 1998 paper by Tom Segalstad at the University of Oslo.[12] Segalstad wrote that 'IPCC's Greenhouse Effect Global Warming dogma rests on invalid presumptions and a rejectable carbon cycle modelling which simply refutes reality, like the existence of carbonated beer or soda "pop" as we know it.' No, I do not understand that either, but Carter apparently does. He, like Segalstad, states that there have been wide fluctuations in atmospheric CO_2 over the past 150 years, ranging up to 500 ppm, all attributed to (unknown) natural Earth processes. Carter refers to a paper by Ernest Beck, who summarised thousands of pre-1950 CO_2 measurements.[13] One such set of analyses comes from Liege in Belgium, made in 1883–4, and it shows CO_2 levels between 310 and 380 ppm. Carter presents this as the range of natural variation in global atmospheric CO_2. These analyses were of air sampled from a window at the author's university in the centre of a city that was then one of continental Europe's largest steel-making centres. Another set of data accepted by Beck (and Carter) was made in 1939 in the urban area of Giessen, an important military centre in Germany at the time. On the basis of these and many similar near-ground urban analyses[14] of atmospheric CO_2, Carter challenges the evidence from Greenland and Antarctic ice-cores and deep-sea sediment cores that post-ice age global average CO_2 levels never exceeded 300 ppm until 1900 (Chapters 8 and 9). Beck claims that CO_2 levels in the 1940s reached 440 ppm, and that Keeling rejected these data because they did not fit his theory of global warming. The trouble with that claim is that Keeling rejected these early values before he had realised that CO_2 was increasing. Keeling found that unless air samples were taken from a region far from industry and habitation, thereby ensuring the air was well mixed, a measurement would only yield a local value, not a representative value for the regional atmosphere.

There is a more serious problem with Beck's compilation, and with Carter's acceptance of it as valuable evidence in the debate about climate. Figure 11.3, reproduced after Beck's, shows that CO_2 apparently increased from 320 ppm in 1930 to 420 ppm in 1940. Where did this sudden burst of CO_2 come from, and what happened to it so that by 1950 it was back to 320 ppm? Ralph Keeling of the Scripps Institute of Oceanography points out that for the world's air to acquire enough CO_2 to shift the concentration so far so quickly would require the carbonisation of one-third of all land plants.[15] The same could be said about the other two peaks, which took place just before 1830 and 1860. Using techniques not available in the 1950s, and which enabled them to distinguish between CO_2 coming from combustion, respiration and the global background, Pataki, Bowling and Ehrelinger[16] in 2002 measured an average 440 ppm CO_2 during winter in Salt Lake City, in the United States. About 70 ppm of this CO_2 was sourced from combustion. How could Beck and Carter, both scientists, accept those earlier wide fluctuations in atmospheric CO_2 as indicative of global averages and not realise that they must be local phenomena – or wrong?

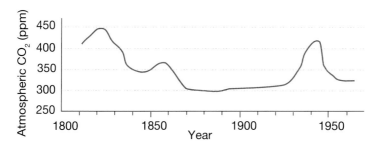

Figure 11.3 Chemically determined atmospheric CO_2, taken from Beck's 2007 paper. Strong increases around 1825, 1860 and 1940 are not detected in ice-cores (Chapter 9) nor by other authors.

There are other late-19th and early 20th-century results Carter might have used when assessing the past 150 years of CO_2 measurements. Callendar was aware of the problems of urban contamination of the air as were Brown and Escombe, who worked at Kew Gardens in London.[17] They reported only values taken when the winds were blowing in from the Atlantic, or from places where there was little industrial CO_2. From 1870 to 1956, according to Fonselius and colleagues from the Swedish Institute of Meteorology in Stockholm, there was a gradual increase in atmospheric CO_2 from about 285 to 320 ppm with a range at any time not much more than 10 ppm. They had set up a series of CO_2 measuring stations in 1954, and from these found a Scandinavian average of 329 ppm over the following year.[18] Fonselius and colleagues were aware of the data Beck was to collate 50 years later, and evidently disregarded many of them.

Carter also plays down the carbon isotope evidence that the increase in atmospheric CO_2 since the Industrial Revolution is a consequence of burning fossil fuels. Recall from Chapter 10 that the signature of coal and oil carbon is a low ratio of heavy to light carbon atoms; that is, a low $^{13}C/^{12}C$ ratio compared to the CO_2 in the atmosphere. When coal burns, it adds proportionately more ^{12}C to the atmosphere than ^{13}C, and so the ratio declines. Carter writes: '... soil and forest carbon, like fuel carbon, are strongly depleted in ^{13}C and therefore feasible sources that may be partly driving the $^{13}C/^{12}C$ decline'. But if you think about this, you have to ask: 'where did the plants and soil organisms get their CO_2?' Out of the atmosphere, did they not? And in so doing they extracted proportionately *more* ^{12}C, thereby enriching the atmosphere in ^{13}C. When they died and rotted (within days if bacteria, or decades if plants, or centuries if trees), the plants returned their carbon, neatly balancing what they took out during their life. Biotic carbon exchange is therefore a zero-sum game and cannot be a feasible source driving the $^{13}C/^{12}C$ decline.

Carter has convinced himself that the contemporary rise in atmospheric CO_2 is largely sourced from the ocean. Being himself an oceanographer, Carter will know that atmospheric CO_2 rose sharply at the end of the last glaciation (from 190 ppm to 265 ppm; see Chapter 8) as the ocean warmed, reaching about 280 ppm 1000 years ago, and had been fairly steadily and slowly declining until the year 1700 (Chapter 9) as the world cooled by almost 2°C. Yet, on page 82 of his book, Carter claims that the increase in CO_2 over the past 200 years comes from the warming ocean:

> These studies imply that the natural processes of marine outgassing of carbon dioxide with temperature rise and juvenile* outgassing from volcanic sources must be much more important, and the burning of fossil fuels less important, in contributing to the current rise in carbon dioxide than is argued by the IPCC.

So here he says, without explanation, that the ocean temperature is rising. Later, on page 147, he states that 'ocean temperatures off southern Australia have been falling for the last several thousand years', and 'Shallow ocean temperatures have been declining since 2004'. But back on page 83 Carter reminds us that ocean degassing of CO_2 follows temperature rise by 1000 years. I find his arguments difficult to follow.

Next, Carter attempts to debunk the greenhouse influence of increasing CO_2 by pointing to two instances where rise in temperature precedes rise in CO_2 concentration. He asks: 'Since when does an effect precede its alleged cause?' His first instance is the rise in atmospheric CO_2 many centuries after the temperature rise following the last ice age. This I explained in Chapter 9, and is neither the subject of climate science controversy nor relevant to the point. Yes, warming the ocean releases some of its dissolved CO_2. That does not stop the released CO_2 from acting as a greenhouse gas and further warming the Earth. Second, Carter notes that the warmth of summer is followed by an increase in atmospheric CO_2 in the winter – first comes the warming, CO_2 rise follows. This is the annual fluctuation in CO_2 as deciduous forests and grasses go through their annual cycle; taking up CO_2 in spring and summer as their green leaves grow, returning it in autumn and winter as the leaves die and rot. Neither of these correct observations has a bearing on the physics of CO_2's greenhouse effect, and it is very surprising that Carter does not understand this.

---

* In geology, 'juvenile' means newly issued from the earth, in this case by volcanoes.

Early in his discussion of CO_2, Carter concludes that only 0.45 per cent of temperature rise is produced by human-derived CO_2; all the rest comes from Earth's natural processes. This amazing and, if true, vitally important conclusion is arrived at like this: If you inject a cubic metre of CO_2 into the atmosphere (1000 litres), after 5–10 years almost all of those particular molecules of CO_2 will have been taken up by plants or micro-organisms or by the ocean. Therefore, there is hardly any of that fuel-sourced CO_2 left to warm the planet, so why worry? Carter seems to have forgotten again the other side of the carbon cycle – plants and organisms die and their carbon returns to the atmosphere. The land and ocean return to the air almost all of the CO_2 they absorb each year. So it may be true that most of that original cubic metre of CO_2 is gone, but approximately half a cubic metre has appeared, indeed by the natural processes Carter is so insistent upon. We add 1000 litres, and after a year there is still 500 litres more CO_2 than there was before, not the 4.5 litres Carter asserts. CO_2 absorbs heat regardless of where it comes from. Plimer's Question 30, explained above, is on the same topic.

The third sceptic view I will comment on is that of Joanne Nova. Nova has a website where she considers, in her words, the only four points that matter.

> 1) The greenhouse signature is missing. This is a knock-out blow.

The 'greenhouse signature' Nova refers to is a prediction based on how greenhouse gases keep the Earth warm. As CO_2 levels increase, the atmosphere absorbs more of the Earth's radiant energy, causing it to warm, with strongest warming in theory over the tropics and between 5 and 15 km high – roughly where aeroplanes fly. Because of the extra absorption at this level in the upper parts of the lower atmosphere (the troposphere), less heat reaches the upper atmosphere (the stratosphere), so according to theory the stratosphere should get cooler. Nova then declares that there is no evidence of maximum tropical atmospheric warming around 10 kilometres, the so-called 'hot-spot', and therefore global warming is not being caused by greenhouse gases.

It is surprising that Nova chooses this to be her number one 'killer' of the theory of global warming by increased CO_2. First, she is accepting a computer model as the test. Here is her stated view of computer climate models: 'But even if they could predict the climate correctly (they can't), even if they were based on solid proven theories (they aren't), they still wouldn't count as evidence.' Then Nova herself uses a computer model as her number one 'knockout blow'.

Second, the set of measurements used to deny the presence of a tropical 10-kilometre 'hot spot' is probably the most controversial set of measurements in all of climate science. Over the past 40 years, the data have been collected from balloons, from satellite measurements and by aeroplanes, and fitting them all together is proving to be a major headache. Some climatologists find no evidence for a hot-spot, while others do. A recent summary and analysis concluded:[19] 'There is no longer a serious and fundamental discrepancy between modelled and observed trends in tropical lapse rates' (temperature change with altitude), while conceding: 'We may never completely reconcile the divergent observational estimates of temperature changes in the tropical troposphere. We lack the unimpeachable observational records necessary for this task.' The important point is that, though climatologists are fully aware of the difficulties in measuring temperatures through the atmosphere, they all find that the lower atmosphere is warming and the stratosphere is cooling. Exactly as greenhouse gas theory predicts.

> 2) Ice cores reveal that CO_2 levels rise and fall hundreds of years after temperature changes.

I have covered this point in Chapter 8 and again in my comments above on Professor Carter's ideas. However, it keeps reappearing in denialists' writings as though they have discovered something the climate scientists do not know. They seem to forget that it was climate scientists who pointed this out.

> 3) The world is not warming any more … true that in the last decade we have had six (or seven or eight) of the top ten hottest years, BUT THAT DOESN'T MEAN MUCH [my emphasis].

It is difficult to think of anything to say to such a ridiculous claim. Every measurement of every kind of temperature except that of the stratosphere shows continuous warming since 1970, and the stratosphere is cooling *because* the lower atmosphere is warming, as explained in point 1 above.

> 4) Carbon dioxide is already absorbing all it can.

No, it is not (see Chapter 4). This is such an unscientific statement, and surprising to read from a graduate in science. Doubling the atmospheric CO_2 content will increase the global temperature by at least 3°C.

Not Plimer, not Carter, not Nova, three of the most vigorous Australian opponents of the idea that climate change can be caused by the burning of fossil fuels, provide one credible scientific paper of direct relevance to support their position. Instead, they present poorly considered counter ideas with no attempt at a coherent alternative interpretation employing physical theory and with no contribution of new observational data. They also miss the point I made at the end of Chapter 8: climate change has several drivers, it can be driven by the sun, it can be driven by the Earth, and it can be driven by us.

Knowing now that climate change is happening, that our burning of coal, oil and gas has been the main driver, and having already seen some of the consequences, what can we expect in the future?

Can we do anything to reduce or remove the excess CO_2?

In the final chapter we will look at how climate scientists are trying to work out what is going to happen to the climate over the next 100 years, and if having pushed it one way, we have any hope of pushing it back.

FURTHER READING

Hamilton C (2010) *Requiem For a Species: Why We Resist the Truth About Climate Change.* Allen & Unwin.

Washington H & Cook J (2011) *Climate Change Denial: Heads in the Sand.* Earthscan.

12

BET YOUR GRAND-CHILDREN'S LIVES ON IT, TOO?

… that's as sure as we ever are about anything. We believe it enough to act as though it is true. When we're that sure we call it knowledge. Facts. We bet our life on it.

Orson Scott Card

The evidence I have presented in this book tells me the climate IS changing, that it is changing in the direction of becoming warmer, that in some
parts of the world it is becoming drier, while in others it is becoming
wetter. It is also evident to me that the reason the climate is changing is
not because the Sun is getting hotter but because we have changed the
composition of the atmosphere by adding CO_2.

It is true, as geologists point out, that the Earth is accustomed to
climate change. Temperatures have been far warmer than they are now;
atmospheric CO_2 has been far higher than it is now, though not during
the past 450 000 years. In 450 000 years there is time for 450 000 generations of most insects, 200 000 generations of most animals and birds,
45 000 generations of most plants and 12 000 or so generations of people.
The climate changes of those long years of successive glaciations were
slow. Organisms had time to move, to remain within their comfort zones,
or to evolve in order to survive.

The common snow gum grows high in the Australian Alps (Figure 12.1),
where it snows every winter and the average temperature is about 6°C
cooler than it is on the mountain slopes 50 kilometres away. In the depth of
the last glaciation, the slopes, one might guess, had an alpine climate, with
snow gums all around. Over a period of 10 000 years, the snow gum seeds
need to blow only 5 metres a year uphill to colonise the mountain tops.

Figure 12.1 Snow gums in the Australian high country are likely candidates for extinction.
Source: James Morrow.

The real problem for the snow gums would be if the high country climate became as warm as the slopes are today. There are no more mountains for the snow gums to migrate up, and they might just become extinct.

In the climate change debate the emphasis is always on global warming and temperature change, and for very good reason. Right at the beginning of this book the Sun was seen to be the driver of climate. The world's climate is a heat engine and the temperature controls it. If the temperature remains the same, the climate stays the same, except perhaps for changes in the paths of ocean current due to the slow, slow drifting of continents. If the temperature changes, the climate changes.

PREDICTIONS

Should the temperature continue to rise at its present rate, it will be doing so at an unprecedented rate in Earth's history. After the Permian glaciation ended 280 million years ago, the temperature rose perhaps 8°C over the following 20 million years, although it is quite possible that there were considerably faster fluctuations. The geological record of so long ago is not as complete as the Antarctic ice-cores, so it is not yet possible to be certain. During the past 450 000 years of Earth history recorded in those ice-cores, temperature fluctuations were wilder: cooling at 1°C in 2000 years and warming at about 1°C in 700 years. Since 1965 the rate has been 1°C in 60 years, 10 times as fast. This is too fast for evolution to allow most species to adapt; too fast for plants to migrate to cooler places by successive generations of seed dispersal.

Is it right to assume that the temperature rise of the past 40 years will continue? After all, there was a decline between 1940 and 1965 related to both industrial smog and perhaps the Atlantic Multidecadal Oscillation (Chapter 3), so another may be due. If you are relying on the Sun to dim, do not hold your breath as the sunspots are again increasing in number towards their predicted maximum in 2013. The Sun's irradiance will increase in parallel. Another Mt Pinatubo eruption, or a bigger one, might cool things for a while, but after the dust settles the CO_2 blanket will still be there to trap the heat.

An important test of a scientific theory is that it should be able to be used to predict. The theory of continental drift predicted that South America and Africa were moving apart. We can now measure their motions and the theory's predictions have been found to hold true. Einstein's famous equation $E=Mc^2$ was theoretical; it correctly predicted that matter could be converted to energy. So it is fair to ask: Are the theories of climate science any good at predicting?

They are. The predictions – 'projections' they are called – are done with computer programs designed to model the behaviour of the climate under various possible changes to the conditions of the Earth and atmosphere. They are complex and sophisticated for they must take into account the wind circulation, the composition of the atmosphere, the Sun, volcanic activity, ocean temperature and current variation with latitude and depth, and many, many more variables.[1] A good climate model is first tested by predicting the past – 'hindcasting'. If it can do this correctly, then its projection into the future might have some usefulness.

Mike Lockwood is a professor at Reading University and at the Rutherford Appleton Laboratory in the United Kingdom. He and his colleagues have used satellite data in particular to measure the variation in the Sun's irradiance since 1975. Working with information collected over the past 50 years, Lockwood has combined four factors known to affect the Earth's temperature in order to 'predict' global temperatures over that period.[2] The factors are:

- the Sun's average radiance
- the known atmospheric dust factors including volcanic eruptions
- the *El Niño* phenomenon
- increasing levels of greenhouse gases over those years.

Figure 12.2 shows how closely scientists are able to 'hindcast' the temperature. Of the four factors used, only the greenhouse gas factor is fitted to the observed increase, in the form of a uniform rise. Such a good match gives some confidence that the climate scientists at least know the main factors affecting global temperatures.

Hindcasting is all well and good, but as the saying goes, everyone has 20–20 vision in hindsight. Twenty years have passed since the first global temperature forecasts were made, long enough to allow us to make an assessment of their accuracy. Since there is not much difference between the model predictions on temperature, I will let one example speak for all.

In 1981, a group of atmospheric physicists at the NASA Institute for Space Studies, led by James Hansen, one who has done more than most to discover the reality of climate change, predicted the degree of global warming to be expected from the rise in atmospheric CO_2. Assuming slow growth in emissions, they predicted a temperature rise by 2010 of

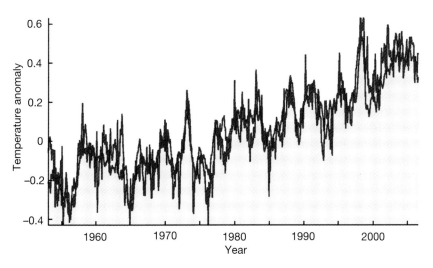

Figure 12.2 The Lockwood match between the observed temperature change (blue line) and the four influencing factors (red line). Source: Lockwood (2008) Figure 2.

0.4°C above the 1950–80 mean.[3] The actual rise was 0.6°C; Hansen and his team were conservative in their calculations. By 1988 there was better understanding of the greenhouse effect, and Hansen, then the leader of NASA's Goddard Institute for Space Studies, made temperature projections based on three scenarios:[4]

- A: rapid emissions growth, no volcanic eruptions
- B: modest emissions growth and one large volcanic eruption in the mid-1990s
- C: modest growth at first, one volcanic eruption, then a reduction in CO_2 emissions by the turn of the 20th century.

Of the three scenarios, B closely matched what happened; the growth in emissions from 1992 to 2000 was modest, and in 1991 Mt Pinatubo erupted. Hansen's 1988 prediction for this scenario put the global temperature for the year 2005 at 0.33°C above the 1988 temperature, and he was close. The increase was 0.34°C. In 2006 Hansen himself said that the close agreement was 'accidental', by which he meant that because there are considerable fluctuations in global temperature one should not expect an exact agreement with prediction in any particular year.[5] He also pointed out that 17 years was not really a long enough period to judge how good the models were.

Computer models are also used to predict sea-level change. In 1990, the IPCC was cautious. Its predictions up to 2100, like Hansen's, included a wide range of possibilities, but all its various scenarios in 1990 indicated a sea-level rise of 60 millimetres by 2010. This figure was changed in 1992 to 45 millimetres after a downward revision of predicted temperature rise. Then, in the next two IPCC reports (1995 and 2001) the prediction for 2010 was much lower, only 30 millimetres, because they did not have firm data on the amount of melting of the Greenland and Antarctic icecaps, therefore they only included alpine glacier melt. The actual figure as determined by satellite measurements is 60 millimetres, for as we now know the icecaps *are* melting. The later, lower estimates made by the IPCC show the conservative nature of climate science's predictions, and are also a reminder that more science is needed to be done to fully understand the factors that make up climate.[6]

The success of those early climate predictions assures us that the scientists know what they are talking about. Contrary to some opinions at the time, the predictions of 1990 were not alarmist; they were realistic. Their accuracy gives us confidence that the longer-term projections of temperature, sea level and rainfall are robust.

CLIMATE SENSITIVITY

Climate sensitivity is the change in global mean near-surface air temperature caused by an arbitrary perturbation in the radiative forcing (see box in Chapter 3) of Earth's radiative balance at the top of the atmosphere with respect to a given reference state.[7] Since CO_2 is such an important amplifier of natural climate change and is a driver of change regardless of how its atmospheric content changes, most of the research on climate sensitivity concentrates on the effect changing CO_2 will have on global temperature. The usual value given is the temperature change that would eventually be produced by doubling the amount of atmospheric CO_2. The sensitivity takes into account the several 'fast' feedbacks that result from temperature change such as from sea-ice cover, water vapour, aerosols and cloud[8] cover. 'Slow' feedbacks, such as changes in vegetation cover and land ice, could conceivably double the values listed below.[9] Knowing the correct value for the climate sensitivity of CO_2 is essential in the construction of meaningful climate projections. It is clear from the following short table that the value is not yet well known.

In its Fourth Assessment Report (2007) the IPCC gave a mean value for this figure of 3°C and a likely range from 2°C to 4.5°C. Other studies yield variations but mostly within this range:

Estimate	Range and source
8.0°	7° to 9° Pagani et al. (2010)[8]
3.0°	or more Royer et al. (2011)[9]
3.0°	1.5° to 4.5° Charney (1979)[10]
3.0°	2° to 4.5° IPCC (2007)
3.0°	2.6° to 3.6° Rojelj et al. (2012)[11]
2.4°	1.4° to 5.2° Köhler et al. (2009)[12]
2.3°	1.7° to 2.6° Schmittner et al. (2011)[7]
0.7°	0.5° to 1.3° Lindzen & Choi (2011)[13]

Because of feedback, sensitivity changes not only from CO_2 changes. For example, if there has been a lot of ice cover in an ice age, as the planet warms there will be a large feedback from shrinking ice. According to Dana Royer and colleagues, whose article provides a summary of this topic, 'a growing body of evidence supports an Earth-system climate sensitivity exceeding 6°C during glacial times and at least 3°C during non-glacial times'.[9]

The most recent IPCC report was published in 2007 and, like the earlier ones, it described different scenarios. If there were no global climate policy, meaning we continue with ever-increasing burning of fossil fuels, we could expect a temperature rise of the order of 3.5° by 2100. Under their most optimistic scenario, which involves major cuts to CO_2 production, the temperature 'only' rises 2°C. I say 'only' because this is the maximum rise that scientists think can be tolerated by the planet without causing irreversible changes, such as complete loss of the polar icecaps.[14]

The scenarios allow projected temperature changes in different parts of the globe to be mapped out (see Figure 12.3, p. 193). They show that in all cases the polar regions warm the most, with a rise of 4–5°C by 2100 in the most optimistic case, almost 8°C under 'no climate policy'. The projections also show that the landmasses warm more than the oceans, and by as much as twice the global average. During the heat waves of the Australian summers of 2008–9 and 2009–10, maxima were between 44 and 45°C. Add 2°C to the global average, and double that for Australia, and the heatwave temperature edges toward 50°C.

As you can see from the numbers in the box 'Climate Sensitivity', not all scientists agree with these projections. Of all those who argue that current climate change is neither abnormal nor significant, the most respected would probably be Richard Lindzen, Professor of Meteorology at the Massachusetts Institute of Technology. Lindzen has produced 240 scientific publications since 1965, with many since 1990 on the topic of global warming and climate change. I think it is fair to say that he considers the variation in global temperature through the past century as having been influenced by increases in greenhouse gases, but not as much as is generally accepted by, for example, the IPCC scientific assessors. Lindzen would argue that the extent and variation in ocean-atmosphere heat exchange (such as *El Niñõ* and the Atlantic Multidecadal Oscillation) are neither satisfactorily understood nor well known. He is particularly concerned that the impact of clouds on the global heat balance is far from adequately understood. These factors lead him to conclude that:

> Using basic theory, modeling results and observations, we can reasonably bound the anthropogenic contributions to surface warming since 1979 to a third of the observed warming, leading to a climate sensitivity too small to offer any significant measure of alarm – assuming current observed surface and tropospheric trends and model depictions of greenhouse warming are correct.[15]

The projections also look at rainfall. Total global precipitation is found to increase in all the models, between 3 per cent and 5 per cent. This is because warmer air can evaporate water from the ocean faster than cold air, and contains a higher amount of water vapour; thus, more water vapour, more rain. Regionally, the tropics get more rain, the mid-latitudes less and the polar regions get more snow.

The IPCC Fourth Assessment looked at a number of other climate changes that can be predicted over the coming 100 years (and beyond). Each projection has a most likely figure with a probable range either side. It does not seem to me to matter too much which particular figure you think might be realistic, since all of them point in the same direction and none of them leave our world unchanged. One of the important uncertainties in the projections arises because we have insufficient knowledge of how the Earth's carbon cycle will react to increased atmospheric CO_2. There is unanimous agreement that the Earth's ability to absorb CO_2 will decrease, but scientists cannot tell precisely how much that will change. Uncertainties such as these indicate why the projections have a fairly wide range of likelihood.

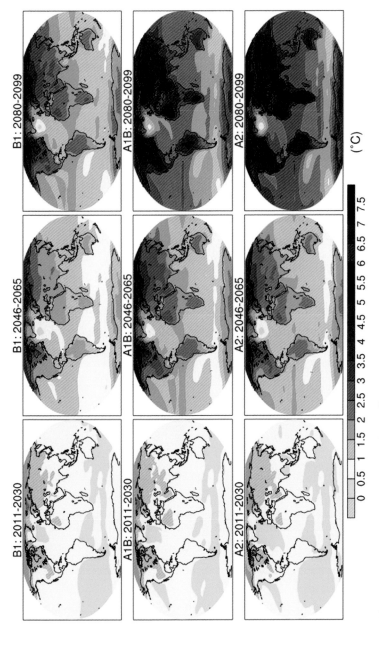

Figure 12.3 Temperature change over three different 20-year periods, projected using the three scenarios: B1 – marked decrease in use of fossil fuels after 2050; A1B: some decrease in use of fossil fuels after 2050; A2: no change from the current trend in fossil-fuel use. Anomalies are relative to the average for the period 1960–99. Source: IPCC Fourth Assessment Report, Figure 10.8.

For the immediate future, say the next 20 years, we can expect small changes:

Temperature rise	at least 0.4°C by 2030
Heat waves	longer and more intense
Rainfall	globally higher; in eastern Australia lower, more intense
Droughts	more frequent
El Niño	no evident change
Sea-level rise	a further 6 cm
Glaciers	decrease, many vanish
Sea-ice	decreases, possibly vanishes in the Arctic
Antarctic icecap	thicker from more snow, continued melting
Sea-level acidity	slight pH decrease, i.e. more acid
Hurricanes	stronger and wetter

On the face of it, all that does not seem too bad, but there are some further serious implications for our grandchildren and their children. If we all sit on our hands, if greenhouse gas emissions simply continue to increase as they have for 50 years, by the year 2100:

- Australia will be 4° hotter

- The Arctic will be 8° hotter

- Sea level will be at least 30 cm higher; some say a metre is not too fanciful a prediction.

Beyond 2100, prediction becomes increasingly uncertain. In building climate models there are many factors the scientists do know about. They know broadly how much warming a given amount of atmospheric CO_2 will cause. They can calculate how much extra water vapour will be put into the atmosphere by that amount of warning and the feedback effect this will have on temperature. They can account for changes in the Earth's albedo and much else.

Then as the saying goes, there are 'known unknowns'. We can guess that sometime in the future there may be a large volcanic eruption that will put dust into the stratosphere and cause some cooling. We just do not know when. We know methane is leaking out of Arctic sediments, but we have no idea how much or how the rate of leakage depends on the ocean temperature. As recently as 1995, the Atlantic Multidecadal Oscillation was known as little more than an uncertain, long-term variable in ocean temperature.

In his book *Storms of My Grandchildren* (2010), James Hansen suggests that climate sensitivity might be changing from a 'known' to a 'known unknown'. We think we know from geological studies of the past 100 million years the extent to which a given increase in CO_2 increases the global temperature (see the Climate Sensitivity box, above). Based on that knowledge, climate projections suggest we can limit global warming if we keep CO_2 levels by 2050 down to 450 ppm. Hansen is far from certain. He suspects, and there is evidence, that the temperature increase per unit increase in CO_2 might have been underestimated by a factor of two (see also Hansen et al. and US Senate Commission on Energy and Natural Resources[4]). Where we now think the world could manage an increase of a further 100 ppm CO_2, Hansen fears we already are close to the limit beyond which the climate could go runaway greenhouse. Remember that Hansen got it right in 1988.

As we saw in considering the carbon cycle, some of our CO_2 emissions are taken up by the ocean. A recent study by Jeffery Park at Yale University in the United States concludes: 'our hypothesis implies that human activities have lately outpaced the ocean's capacity for absorbing carbon', leading to the suggestion that its ability to absorb CO_2 might be declining.[16] If so, the temperature projections will all be too low.

And there are the 'unknown unknowns'. Twenty years ago, the climate feedback from methane emissions were in that category; who knows what other surprises await us as the warming trend continues?

If the levels of greenhouse gases were to be stabilised today, there would still be at least a further 0.5°C of warming thereafter.[17] Stabilisation is not likely to happen soon, so that 0.5°C is on top of whatever rise occurs before we stop increasing greenhouse-gas emissions. Once their levels start to decrease, it would still take centuries before global temperature would begin to fall. However, even if emissions stopped altogether by 2100, it would take 1000 years or longer for the climate to settle down, and it would then still be a couple of degrees above pre-industrial levels.[18]

TIPPING POINTS

There is a fear that the modest temperature and climate changes outlined above are too optimistic. The fear is that the climate system has built-in instability about which we know too little, and that a small change can tip the scales.

In most of the scenarios modelled by climate scientists, there is the presumption that we will be smart enough to reduce our CO_2 emissions some time before 2050, sufficiently to keep the overall temperature increase to no more than 2°C above pre-industrial levels. But we do not really know if that will be possible. It could be that another 1°C rise might be enough to warm the Arctic and Antarctic sea-floor sediments sufficiently to release a huge burst of methane. Methane is about 30 times as strong as CO_2 at warming the globe. There is more carbon in the methane stores than in all the fossil fuel stores put together. If that came out quickly, 2°C would be blown away as the temperature roared past.

Melting Arctic ice is another potential tipping point. There is worrying evidence that the ice melt is accelerating, not just warmed by the warming atmosphere but also from below by the warming ocean, and that melt water is lubricating the glaciers and accelerating their slide into the ocean. Lose too much ice and we lose too much solar reflection, thereby accelerating the warming. Recall what happened when the Atlantic sea-ice melted at the end of the Younger Dryas (Chapter 9). The Greenland air temperature rose very quickly, several degrees in a few decades. Could such a rapid warming occur over the Arctic if the sea-ice cover is lost?

The great ocean currents could be another tipping point. Some oceanographers suspect that another degree rise in temperature could tip the Pacific Ocean into a state of persistent *El Niño*. Changing the Arctic surface waters could affect the essential 'conveyor belt' that stirs the whole world's oceans.[19] Cold Arctic water is denser than warm, saltier tropical water. In the north Atlantic, these density differences drive a current northward, and because seawater density is determined by temperature and salinity, the resulting circulation is termed 'thermohaline'. The cold Arctic water sinks and flows south, to be replaced by the warmer, tropical near-surface water, largely moved by the wind – the Gulf Stream. If much freshwater were to enter the north Atlantic, say from the melting of the Greenland icecap, this less-dense addition to the ocean could slow the sinking of the Arctic water and in turn reduce or stop the existing system of currents. We saw in Chapter 7 that it looks as though we are close to a tipping point for some marine creatures, such as the tiny plankton that feed so much of the ocean's larger animals. Their demise could upset the entire ocean ecosystem.

A tipping point for humans is the point we really do not want to reach. Steven Sherwood from the University of New South Wales and his colleague from the United States, Matthew Huber, discussed the likely effect on humans of a temperature rise of 7°C.[20] At present there is nowhere on Earth where the temperature gets so high that humans cannot keep cool; that is, keep our core temperature at 37°C. It might be 45°C in the shade, but evaporation keeps the skin to about 35°C and that is enough to keep us cool. Generally, such high temperatures are associated with low humidity, so evaporative cooling works well for us. Where humidity is very high, as in the tropics, the temperatures are rarely over 35°C, anyway. A 7°C rise could happen by about the year 2150 if we continue to burn fossil fuels at the presently increasing rate, and then we will see some regions of the Earth pass the limit of human endurance. At such elevated temperatures you cannot cool down, and heat stress quickly proves fatal.

If you think this prediction is fanciful, it is worth reading the authors' words about their analysis, for they show once again that scientists are by nature cautious rather than alarmist. They write: 'Our limit applies to a person out of the sun, in gale-force winds, doused with water, wearing no clothing, and not working.'

IMPLICATIONS

Some of the effects of climate on plants and animals are already obvious, such as those I discussed in Chapter 2. Earlier harvests and changed migration patterns do not at first glance appear to be very significant. Warmer temperatures might bring on earlier crops, but not necessarily better ones. In a study of wheat growth in northern India, it was found that at temperatures above 34°C, wheat reaches the end of its growth before the grain is filled, thus reducing crop yield.[21] There are many other changes we already know about, and changes in one thing can cascade to larger changes across the environment. One of the favourite arguments of those who deny the reality of climate change is that increased atmospheric CO_2 results in increased growth rates and increased leaf mass for most plants that have been studied. It is claimed that CO_2 is therefore good for the environment. Such results have certainly been seen in small laboratory experiments; however, the results are less spectacular in open-field studies of plots bigger than 100 square metres. Rice and soybean, two of the world's most important crops, show a small increase in harvestable yield under an atmosphere of more than 600 ppm CO_2.[22] In summarising 30 years of experimental studies, known as Free Air CO_2

Enrichment, where the CO_2 levels were elevated across large field trials, Andrew Leakey and colleagues at the University of Illinois in the United States concluded that under enhanced CO_2 levels crop yield increase was small, of the order of 15 per cent.[23] An earlier study had found that young trees did better than legumes and grasses responded worst.[24] Another advantage of high CO_2 is that plants reduce the size of the pores that exchange gases with the atmosphere, and in so doing they reduce their rate of water loss and become more drought tolerant.[25]

There is unfortunately a down side that tends to counterbalance the extra growth and reduced water loss. CO_2-enhanced growth results in plants having lower nutritional value than normal plants. Wheat and rice both have lower nitrogen content, and leaves become less nutritious for grazing animals. A Japanese study found that there was about 13 per cent decline in the nutritional value of rice grown under enhanced CO_2, though largely compensated by a higher yield.[26] Much the same conclusion was reached in an analysis of more than 200 experiments on barley, rice, wheat, soybean and potato: the grains and potato lost 10–15 per cent of their protein value, whereas soybean was little affected.[27] According to Petra Högy and Andreas Fangmeier, for wheat the impact of rising CO_2 on crop quality will be small but, 'The amount of protein in wheat grains and products derived from wheat will be reduced in a CO_2-rich world.'[28] There is more on the down side; in the experiments at the University of Illinois using artificially heightened CO_2 levels described earlier, the enhanced growth was somewhat negated by increased attack by beetles.[20]

Possibly the decrease in food value might be offset by improved strains of crop plants and increased areas of arable land. For southern Australia, the predicted temperature increase of as much as 4°C by 2100 will probably be coupled with lower rainfall and more frequent drought, and that will seriously contract the area available for farming. Increased rainfall in the tropical north will not provide a balance; there is already enough rain in the wet season and CSIRO studies of the potential of the area for significantly increased agriculture were highly pessimistic.[29] It is often suggested that northern Australia could support a magnificent rice industry. The CSIRO report specifically addressed this suggestion and classified almost the entire area as 'marginal or unsuitable' for rice growing.

Australia is one of the most fire-prone continents, though the character of wildfire, known as bushfire to Australians, varies considerably. In the tropical north, grass fires are set every year by graziers as a way to increase stock fodder. Each such fire is quite small, but overall much of the 'top end' of the country may be burnt in a year.[30] While the frequency of the fires deliberately lit bears no relation to climate, their extent may be.

It is commonly said that the Australian bush benefits from burning. From the perspective of the northern Australian beef industry, this is probably true. Frequent burning reduces the variety of plants, leaving more open space for grasses to grow. If fires pass through more frequently than the time it takes a tree or shrub to mature and set seed, that species will die out, so will the insects that depend on it and so will the lizards and birds that eat the insects, and so on.

Unlike wattles and gum trees, Little Penguins nesting among the tussock grass on Victoria's Phillip Island did not evolve with fire. It seems they have no instinctive avoidance of grass fires, and simply sit in their burrows until they are burnt alive.[31] In Victoria in particular, the landscape is covered by either forest or farmland. Fire in Victoria is always a danger and always a disaster if it gets out of control. Severe bushfires occur, almost without exception, at the height of summer, after a period of high temperatures and generally under the driving force of a hot northerly wind. It is plausible at least that either from rising temperature or declining rainfall, climate change could be a factor in changing the extent and frequency of such bushfires.

Following disastrous fires in 2009, the State of Victoria held a Royal Commission of Inquiry, which among other things listed the major bushfires that had occurred in Victoria since 1850.[32] One might imagine that as the years passed, decreasing areas of forest, increasingly improved access to the bush and improvements in methods and machinery for fire-fighting would have led to improved fire control and a decline in the extent of the area burnt. Or one might suppose that an increasing population and an inevitable increase in the number of potential fire starters (accidental or deliberate) might have had the opposite effect.

Whichever of these three factors, climate, fire control or people, have contributed to the incidence of Victorian bushfires, the evidence seems to indicate that there was an increase in severity around 1940, and again in the past 10 years. Figure 12.4 (p. 200) shows the cumulative area burnt from 1895 and gives some idea of how the fire extent has changed. Other than an immense, almost statewide bushfire in 1851, few major fires occurred before 1930, then there were three large conflagrations, in 1939, 1944 and 1952. Coincidentally, perhaps, these occurred at the end of the marked rise in global temperatures from 1920 to 1945. Things seemed to settle down over the next 45 years, and for the first part of that period so did the rise in global temperature. But from the end of the 20th century,

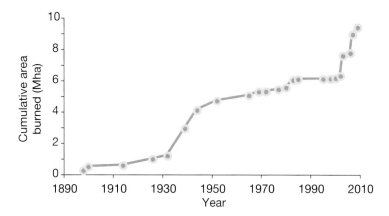

Figure 12.4 Cumulative area burnt by bushfire in Victoria from 1896. Gentle slopes in this graph from 1896 to 1930, and again from 1955 to 2000, suggest more-or-less average climatic conditions. Sharp rises in fire extent from 1939 to 1952, and again after 2000, might reflect rising temperatures in the preceding years.

and after the sharp rise in global temperature from 1975 to 2000, Victoria experienced four extreme bushfires.

One of the predictions made by climate scientists is that as the global temperature rises, so will the incidence and severity of wildfires. Are we seeing this prediction coming true in Victoria's bushfires?

Biologists have predicted that some species will not be able to adapt to climate change, and the evidence that they are right is beginning to be uncovered. A study of lizard populations across four continents by 26 scientists from 20 research institutions found that as a species, these animals are suffering heat stress.[33] In the spring, when they usually breed and are in most need of food, it is becoming too hot for prolonged searching, and they starve as a result. Since 1975, 12 per cent of the lizards in local Mexican populations have become extinct, and 21 per cent in Madagascan nature reserves. The study's last words were: 'Our findings indicate that lizards have already crossed a threshold for extinctions.'

Rising sea levels feature prominently among the known effects of global warming. As we saw in Chapter 7, sea level is now rising at about 3 millimetres per year. If polar icecaps are indeed melting at the speed some studies suggest, that rate is set to increase. James Hansen regards sea-level rise as one of the 'dangerous' consequences of global warming.[34] If the icecaps do melt significantly, sea-level rise will be measured in metres, not millimetres, and most coastal cities will be unsustainable. Professor Stefan Rahmstorf of the Potsdam Institute for Climate Impact

Research in Germany estimates that a realistic figure for sea-level rise by 2100 is a little over 1 metre.[35] That leaves 64 metres to go if all the polar ice melts and we return to the Cretaceous shoreline of Figure 7.1a. CSIRO scientists estimate that a sea-level rise of half a metre will increase the frequency of coastal flooding from storms in the Cairns, Queensland, region by a factor of almost three.[36]

The Great Barrier Reef is worth $40 billion, according to a study commissioned by Great Barrier Reef Foundation.[37] That is the Foundation's estimate of the loss to the Australian economy should the reef die. According to those who deny global warming, 1998 was the warmest year on record and the world has been cooling since. The jury is still out about the subsequent cooling, but certainly 1998 was a warm year globally. And in that year, 60 per cent of the Great Barrier Reef experienced coral bleaching. Most of the reef recovered, but 5 per cent did not.[38] It is not expected that the reef will be able to survive repeated episodes of such bleaching. We are currently (2012) under an atmosphere with 396 ppm CO_2 and increasing at 2 ppm annually. By the time we reach 500 ppm CO_2, which could be less than 50 years away, coral reef scientists predict: 'Above this point coral reefs will also change irreversibly and be lost for many thousands of years', and 'coral dominated reefs will be rare or non-existent in the near future.'[39] If cyclone intensity rises, as some predict, damage to the already compromised reefs will accelerate their demise.

So do we bet our grandchildren's lives on the science of climate change? Perhaps that depends on what we value. If we feel that the trees and flowers, the animals and butterflies we grew up with are 'good', that we would like them to stay around for our children's children, then we might conclude it is time to act. Pteropods make up a large fraction of the cold oceanwater plankton; they provide food for larger invertebrates and fish, as much as half the diet of juvenile pink salmon, for example.[40] No pteropods, no salmon steaks. If we do not want our grocery, air-conditioning and water bills to take up a large chunk of our income, and we do not like watching vulnerable people dying from heat prostration, we might think it is time to act.

Or we might not care about those things. Perhaps we accept that what happens is what happens and that humans are as much part of nature as caterpillars, so global warming is just the natural scheme of things. In such a 'do nothing' scenario, here in Australia I expect some people would manage fairly well, though their beach house might be in trouble. We could, for example, use our huge uranium resources to build many nuclear power plants, or put solar panels all over the Simpson Desert, and use the electricity to run coastal desalination plants to provide irrigation water to the parched Murray-Darling basin.

John Schellnhuber offers some worrying thoughts about the failure of the world to tackle climate change, albeit with a humorous touch.[41] For example, he suggests that society's approach to decision making is: 'Do not argue with the scientific mainstream (at least not in public), but ignore its recommendations when it comes to practical purposes.'

Following the December 2009 Climate Conference in Copenhagen, Schellnhuber noted that the national offers for CO_2 reduction revealed that our planet is heading for a medium anthropogenic warming of 3.5°C. Yet climate science warns that 2°C is the most we can allow. According to Schellnhuber, 'the most aggressive global emissions reduction strategy we can possibly get will ensure only a three-in-four chance to hold the 2°C-line'. Worse odds than Russian roulette, he comments, but then 'its only our planet'.

Schellnhuber considers our attitude to global warming in a gambling context. He supposes a game where there is a high probability that unabated global warming will generate dangerous, if not disastrous, impacts and a low probability that humankind will respond with appropriate mitigation and adaptation measures, to the challenge as sketched by science.

The least likely outcome of this game of chance is that science is wrong but the world responds by ceasing fossil-fuel emissions. In that case the scientists look stupid but the world benefits by the extended availability of fossil fuels and the development of new energy technologies.

The most probable outcome is that science is right, nothing is done about it, and disaster follows.

There are two other outcomes, both equally unlikely. One is that science is wrong and nothing is done. Science looks foolish and later, should a real crisis appear, nothing is done again; the 'cry wolf' syndrome. The other is that science is right, and everything is done to mitigate climate change and the world is saved.

What follows from this is that the odds of winning by doing nothing are less than one in 10. Would you bet your grandchildren at odds of 10 to 1?

TIME FOR ACTION

You have found out that we are warming the globe. You know that if we do not stop burning fossil fuels we will change the world we inherited.

What are you going to do about it? So far, except for the last few paragraphs, I have been careful to report quality scientific results, information that can be checked, verified, repeated if you have the knowledge and equipment, and analysed. Now I am going to reach beyond my scientific comfort zone.

ALTERNATIVE ENERGY

Slowing down and eventually reversing the build up of CO_2 and other greenhouse gases in the atmosphere requires people to do something. Perhaps you might buy an electric car and then power it from a coal-burning power station. Petrol produces 20 per cent less carbon per kilowatt than coal; in this case, petrol cars are the better option. Where electricity is generated from other fuels or renewable sources, electric cars win, hands down. According to Wikipedia, electric cars in the United Kingdom save 40 per cent of carbon emissions, compared to petrol and diesel.

In the 1960s, the scientist James Lovelock recognised that the Earth is a huge, complex system, which he called Gaia. His ideas have been instrumental in the thinking of environmentalists and the policies of the Green Movement ever since. In an article published in *The Independent* (24 May 2004), Lovelock wrote:

> We cannot continue drawing energy from fossil fuels and there is no chance that the renewables, wind, tide and water power can provide enough energy and in time. If we had 50 years or more we might make these our main sources. But we do not have 50 years; the Earth is already so disabled by the insidious poison of greenhouse gases that even if we stop all fossil fuel burning immediately, the consequences of what we have already done will last for 1,000 years. Every year that we continue burning carbon makes it worse for our descendants and for civilization.

So what does Lovelock conclude? Go nuclear. He writes: 'Opposition to nuclear energy is based on irrational fear fed by Hollywood-style fiction, the Green lobbies and the media. These fears are unjustified, and nuclear energy from its start in 1952 has proved to be the safest of all energy sources.'

Perhaps, then, if we are opposed to nuclear energy we might want to rethink our position. Nuclear is a known technology; an operating nuclear power station produces no CO_2, and there is plenty of fuel available.[42] This is not the place to discuss the nuclear option at length, but at least the possibility should be considered. A good place to get ideas would be the book *Plentiful Energy*, listed at the end of this chapter.

Wind is one of the best-known and well-developed sources of alternative energy. In Europe, 18 per cent of the total electricity production comes from the wind, with farms like the 102 turbine off-shore Walney Wind Farm in the United Kingdom, which is capable of generating 350 megawatts. In an article that discusses many of the pros and cons of alternative energy, Mark Jacobsen and Mark Delucchi at Stanford University in the United States consider that by 2030, when the global demand would be about 17 000 gigawatts, 4 million 5-megawatt wind turbines occupying 1 per cent of the global land area could contribute 50 per cent of that requirement.[43] But a completely different interpretation is made by some scientists in Germany. Assuming 100 per cent efficiency, they calculate that the maximum extractable energy from the wind, using *all* the Earth's land surface that is not glaciated, is only 37 500 gigawatts.[44] In addition, they estimate that extracting all that wind power would have as much effect on climate as doubling atmospheric CO_2.

At present the best a solar panel powerstation can produce is about 1 megawatt for 1 hectare of panels. This is an area of rapidly advancing technology, and it is not unreasonable to guess that their efficiency in 25 years time might increase from their present 15 per cent. Jacobsen and Delucchi calculate that 20 per cent of the 2030 power requirement could be sourced from solar panels, using rooftop systems for 6 per cent and 300-megawatt plants for the rest.

Clearly, there are differences of opinion in deciding the way forward using alternative energy sources. Jacobsen and Delucchi reject nuclear energy for a variety of reasons, including political and social issues and limited uranium supplies, and their estimate of power demand for the future is lower than some other estimates. British Petroleum's *Statistical Review of World Energy,* published in January 2012, predicts a world demand for energy of 23 000 gigawatts by 2030. According to the United States Energy Information Administration's *International Energy Outlook 2011*, global energy consumption will rise by 53 per cent by 2035, which would put the demand then at 25 000 gigawatts.

Using the last figure, let us look at what is involved to replace the world's fossil-fuel power generators with nuclear, wind and solar by 2035. For the ideas that follow, but not the figures, I am indebted to Andrew Charlton's article in *Quarterly Review,* referenced at the end of this chapter as further reading. To generate 25 000 gigawatts, the world must build for the next 23 years:

- Twenty 1.5 GW nuclear power stations every month, to provide 9,000 GW, AND;
- 800 windfarms every year (that's ten 3.5 MW turbines an hour), for 7,000 GW AND;
- 100 square metres of (new technology) solar panels every second for 9,000 GW.

These are not the only ways to produce alternative energy, but they are the three most commonly discussed, and they serve to make the point that replacing coal, oil and gas is no simple matter.

CARBON SEQUESTRATION

Wally Broecker, a geochemist at Columbia University in New York, was one of the first to recognise the impact of burning fossil fuels.[45] He has very strong views on this subject, views that deserve to be taken seriously, given his vast knowledge and expertise. Broecker accepts that the developing countries of the world are not going to stand by quietly without getting their share in the standard of living that fossil fuels have brought to the rich countries. He argues that fossil-fuel use will inexorably increase, and that atmospheric CO_2 will easily pass 560 ppm, a level most climate scientists regard as already too high. Eventually, though, he is optimistic enough to think that alternative energies, solar or nuclear, will win out.

Broecker does not suggest that wind energy can replace coal and oil. He argues that using the wind is just as much a climate changer as adding CO_2, and one we know nothing about. Wind is air in motion, and air carries water. Wind farms take the energy out of the wind; they slow it down. Slowing enough wind to replace fossil fuels as an energy source could have an enormous influence on rainfall, in particular, and climate in general. Nuclear energy has its own political and social issues, and Broecker suggests that covering large areas of land with solar panels will attract at least as much opposition as wind farms. That leaves, in his opinion, only the removal of CO_2 from power stations as it is produced, or from the air itself, as viable options for this century. This is known as 'carbon sequestration'.

First, the CO_2 has to be extracted from wherever it is, either in a smokestack where it is fairly concentrated, or in the air, where it is very dilute. There are several ways in theory of doing this, and much research is currently devoted to the topic. Some methods involve absorption of the CO_2 by a chemical (just which chemical is a proprietary secret). Others freeze the CO_2 out of the power-station flue gases. In Australia, research into carbon sequestration is being undertaken at a government-funded Cooperative Research Centre for Greenhouse Gas Technologies, known as CO_2CRC, in collaboration with the CSIRO.

Once the CO_2 has been extracted it has to be permanently stored. Only two ways of doing this seem to have any practicality: bury it as high-pressure CO_2 or convert it to a mineral form, say as calcium carbonate, in the way that shell fish do. The burial option is by far the most popular because the technology needed to do this already exists. CO_2 is routinely compressed and piped for industrial use, and injecting it into underground reservoirs is a technology well known to the oil industry; they do exactly that to flush oil out of some source rocks. The Australian CO_2CRC has successfully demonstrated in the Otway Basin of Victoria that compressed CO_2 can be buried in sediments at a depth of 2 kilometres. This approach, which is the same as that which is being researched in the United Kingdom, uses the highly practical option of returning the CO_2 to the underground rocks from which oil has been extracted. Such rocks are able to store gas safely and permanently; otherwise they would not have kept their oil for millions of years. A similar approach was proposed by former US president Bush's now-abandoned FutureGen project, which aimed to develop coal-fired power stations with low-carbon emissions by pumping the CO_2 into underground salt-water rock storages (aquifers).

CO_2 could also be pumped to the bottom of the ocean, where it should sink, though just what it will do to the ocean's chemistry is unknown. My view is that it is risky to tamper with processes we do not fully understand. There are examples of when we have done such things before, like introducing cane toads to Queensland to clean up sugar-cane beetles. This has not worked, and the proliferation of cane toads is considered an environmental disaster across northern Australia.

According to Broecker, too much of our CO_2 emissions occur away from point sources like power stations. Cars, trains and planes, houses, farms and land-clearing are unsuited to the notion of CO_2 capture at source. So Broecker's approach is simple: clean up the air. For thousands of years, humans lived with their own waste products. Streams and ground water became polluted, and resulting sicknesses killed more than half the babies born, well before they reached 12 months. In ignorance we did nothing about it for a long time. Eventually, we discovered that sewers and sewage treatment plants could eliminate the pollution. We learned to clean up our waterways. Broecker says we can now do the same thing to the air and remove the CO_2. He has been active in developing technology to do that, with the result that the company Global Research Technologies has developed an air scrubber. It can remove a tonne of CO_2 a day from the air and, ideally, convert it to an inert mineral form. Broecker says it would take 10 million such scrubbers, each the size of a shipping container, to clean up the air. That is not so many compared to, say, the 600 million cars in the world.

There is another way to remove some CO_2 and methane from the air. Biochar is the name given to a slow-burning product that is similar to charcoal. Charcoal is made by burning wood in such a way that much of it is converted to carbon, with the release of heat when the other components, primarily hydrogen, burn away. In contrast to charcoal, biochar can be made from any organic waste, such as sugar cane, solid sewage, branches and leaves from forestry and so on. Typically, this kind of material is either burnt to ash, with a rapid release of CO_2, or buried, where it is converted in a year or two to CO_2 and methane by soil organisms. Either way, all its carbon ends up back in the atmosphere quite quickly. This is part of the normal carbon cycle.

The value of biochar is three-fold. First, it retains about half of the carbon that would otherwise be released to the air, and when buried it is very slow to be converted to CO_2; it takes over 100 years or so. By burying biochar we could reduce CO_2 emissions from land sources. Second, with suitable technology, making biochar generates heat, which can be used to replace energy sourced from fossil fuels. Third, buried biochar improves soil fertility, something that has been known and used by farmers for thousands of years.

Writing in the journal *Nature Communications*, Dominic Woolf and colleagues[46] demonstrate that at its best, biochar production and storage in soils could reduce our CO_2 outputs by 12 per cent. This would make a very real contribution to reducing increases in atmospheric greenhouse gases while we slowly convert our energy systems.

GEO-ENGINEERING

Pumping CO_2 into the ocean, or perhaps even into rocks, could have unexpected consequences. A proposition for which some of the consequences are reasonably well known has been put by a number of eminent climate scientists, notably Paul Crutzen.[47] Crutzen has saved the world once already; in 1995 he, along with Sherwood Rowland and Mario Molina, were awarded the Nobel Prize for their work explaining how chlorofluorocarbons (CFCs) were depleting stratospheric ozone. The idea Crutzen now espouses is simple: copy Mt Pinatubo. In 1991, this volcano put about 10 million tonnes of sulfur in the form of sulfur dioxide (SO_2) into the atmosphere. That SO_2 was converted to tiny droplets of sulfuric acid (battery acid), which is a strong reflector of light. This caused the sunshine to be dimmed, as we saw in Chapter 3, and the global temperature fell by about 0.1°C. The idea is to spray perhaps 5 million tonnes of sulfur as SO_2 into the stratosphere, where it will be converted to sulfuric acid and reflect sunlight. After about a year the acid will wash down into the troposphere (lower atmosphere) and need to be replaced, so this would be an ongoing piece of geo-engineering. We already put 60 million tonnes of sulfur annually into the troposphere by burning fossil fuels,[48] so we know the side effects: smog, asthma, pneumonia, death. The total annual input of sulfur to the lower atmosphere from volcanoes (7 per cent), biogenic sources (23 per cent) and fossil fuels (70 per cent) is 100 million tones. The additional few per cent is not thought to be very significant.[49]

There are uncertainties about this, but sulfate aerosols might be a way to buy time. Tom Wigley, another leader in the world of climate science, is a graduate of the University of Adelaide in South Australia. He was a director of the Climatic Research Unit at the University of East Anglia in the United Kingdom before moving to the National Center for Atmospheric Research in Boulder, Colorado. Wigley argues that while there is more to the problem of increased CO_2 than just temperature rise, slowing global warming for the next few decades with sulfuric acid would allow the political process of reducing CO_2 emissions to develop to the point at which the sulfate sun-shield was no longer needed.[50] Stanford University scientists Ken Caldeira and Lowell Wood have calculated the overall climate effect of such geo-engineering, and conclude that it would work to limit global temperature rise.[51]

In the journal *Science*, Richard Kerr summarised the pros and cons of this suggestion that we geo-engineer the planet.[52] The main argument against doing this is that it will remove the pressure to reduce CO_2 emissions, and the longer we allow emissions to rise, the worse the situation will become if (or when) the sulfuric acid 'band-aid' is stopped. The CO_2 would still be there, and as soon as the sulfuric acid is all rained out, the temperature would increase in a big hurry. Another concern is that while we know what sulfuric acid in the troposphere does, we have much less understanding of what it will do to the stratosphere. That is where the ozone layer is. Wigley recognises that adding SO_2 to the stratosphere will enhance ozone loss, but regards this risk as small and short term (40 years). There is some irony when Paul Crutzen, the man who won a Nobel Prize for saving the ozone layer, is now championing a method of slowing down its recovery. The sulphuric-acid solution is so cheap that one billionaire could decide to do it all alone, and so could one small nation. It does not require international agreement in the way that carbon reduction does.

Katherine Ricke of Carnegie Mellon University in the United States and her colleagues[53] have modelled the impact on temperature and global rainfall that would be caused by a sulfate sunshield, tactfully called 'solar-radiation management'. They agree that the use of this technology would reduce the extremes of temperature and rainfall, but that both cannot be stabilised together while atmospheric CO_2 increases. Furthermore, achieving stability of climate worldwide would not be possible, so one region might benefit at the expense of another. This is not a good scenario for political quiet.

Other methods of cooling the Earth have been suggested, such as creating a shield in space to block some sunlight, or altering cloud reflectivity by spraying seawater high into the air. All these ideas are put forward by well-meaning, knowledgeable climate scientists, though for none of them is it their preferred action. But they have become resigned, particularly since the failure of the 2009 Copenhagen Conference, to the reality that populations and politicians are not going to reduce carbon emissions any time soon. With that reality, perhaps the worst effects of global warming can be reduced by a sulfuric acid sunshade, while the increasing ocean acidity from rising CO_2 levels, the reduced nutrition of crops and a bit more acid rain will just have to be tolerated.

WHAT CAN YOU DO?

The problem is CO_2, not the Sun. Capture and storage is a sensible proposition, but in the long term it is not enough. We need to reduce our carbon use, and this will only happen if it is economically desirable. This could occur simply by encouraging research into alternative energies until these techniques were able to generate electricity more cheaply than coal, oil and gas. Energy companies follow profits, so they will switch. The trouble is, without some incentives to fast-track the research, it might all be too late. Do not forget the tipping points; they are not far away. Even in war time, it took the United States 6 years to develop an atom bomb. The Montreal protocol on substances that deplete the ozone layer was signed in 1987. Though most are now gone, the last of these gases will not be phased out until 2030. How long will it take to replace coal and oil in the absence of strong political and social pressures? And in the face of the combined resistance of the power industries?

As a start, pushing the energy economy away from reliance upon coal would be a help. Coal currently contributes about 40 per cent of the world's CO_2 emissions. Converting coal-fired power stations to natural gas would reduce CO_2 emissions by almost a half, because natural gas gets some of its energy by converting hydrogen to water, whereas coal only has carbon to work with. But that is in theory. In practice, as Tom Wigley points out, the extraction of natural gas, particularly by fracking, is likely to leak methane to the atmosphere, and its greenhouse-gas effect would substantially cancel the advantages of replacing coal with gas.[54]

But that is only a first step. After gas, what: solar, nuclear, wind, tidal, photovoltaics? A move from fossil fuels to alternatives will be slow, for as we have seen, it would require a vast infrastructure to be built very quickly. That would not even begin without the political, social and industrial will to do so. So *we* need to apply the pressure. In democracies, at least, the people have some power, because governments eventually listen to us in order to get elected. Join a pressure group, form a pressure group, write letters to your local member of parliament, write to the newspapers, convince friends. You need to persuade the government to make the changes happen. But do not delude yourself that a change to carbon-free energy will be free. We have had a free ride on fossil fuels for 200 years, ignorant of the impact we were making. Now that we know the cost, it is time to pay the price. Either you will pay more in taxes, or you will pay more for power and more for the things you buy that use power in their manufacture and delivery. The money can be used to remove CO_2 and on developing other forms of energy. There are plenty of ideas on how it might come about, but you will not read them here because economics and public policy are not scientific disciplines. You do not need my views on those.

SUMMARY

Topic	Observation	Conclusion
Predicting climate	Models match past climate very closely.	CO_2 is the cause of global warming.
'Safe' CO_2 limit	Doubling CO_2 in the atmosphere raises the temperature by about 3°C.	Many feel 450 ppm should be the limit.
Tipping points	Methane release, Arctic ice melt, ocean convection can all change suddenly.	Be prepared.
Implications	Widespread: drought, flood, heat, species decline, agricultural uncertainty may all stem from climate change.	Reduce atmospheric CO_2.
Alternative energies	Too little, too late? Replacing fossil fuels other than with nuclear requires huge, rapid infrastructure development.	Re-assess nuclear and research solar-energy technologies.
Geo-engineering	Carbon removal at source and from the air is possible. Aerosol sunshades work in theory, albeit patchily.	Carbon removal provides a way forward, but it will cost. Sunshields need more investigaton.

FURTHER READING

Bradstock RA, Gill AM & Williams RJ (eds) (2012) *Flammable Australia: Fire Regimes, Biodiversity and Ecosystems in a Changing World*. CSIRO Publishing.

Broecker WS & Kunzig R (2008) *Fixing Climate:What Past Climate Changes Reveal about the Current Threat – And How to Counter It*. Hill & Wang.

Charlton A (2011) 'Man-made world: Choosing between progress and planet. *Quarterly Essay 44*, 1–72.

Cook PJ (2012) *Clean Energy, Climate and Carbon*. CSIRO Publishing.

Goodell J (2010) *How to Cool the Planet*. Houghton Mifflin Harcourt.

Hansen JE & Sato M (2012) 'Paleoclimate implications for human-made climate change.' In Berger, Mesinger & Sijaci (eds) *Climate Change at the Eve of the Second Decade of the Century: Inferences from Paleoclimate and Regional Aspects: Proceedings of Milutin Milankovitch 130th Anniversary Symposium*.

Schneider SH, Rozencrantz A, Mastrandrea MD & Kuntz-Duriseti K (eds) (2010) *Climate Change Science and Policy*, Island Press.

Till CE & Chang YI (2011) *Plentiful Energy – The Story of the Integral Fast Reactor*. Createspace.

NOTES

1 THE SPIRIT OF ENQUIRY

1   Mann ME, Bradley RS & Hughes MK (1998) 'Global-scale temperature patterns and climate forcing over the past six centuries.' *Nature 392*, 779–87.

2   Alex Jones's Infowars.com. (16 December 2009) 'Climate change questions: Andrew Glikson of the Australian National University at Canberra talks with Lord Monckton.' Viewed 22 May 2012 at www.infowars.com/climate-change-questions.

2 GLOBAL WARMING

1   Aono Y & Kazui K (2008) 'Phenological data series of cherry tree flowering in Kyoto, Japan, and its application to reconstruction of springtime temperatures since the 9th century.' *International Journal of Climatology 28*, 905–14.

2   Chuine I, Yiou P, Viovy N, Seguin B, Daux V & Le Roy Ladurie E (2004) 'Grape harvest dates and temperature variations in eastern France since 1370.' *Nature 432*, 289–90.

3   Decanter.com, 31 August 2011.

4   Sparks TH & Menzel A (2002) 'Observed changes in seasons: An overview.' *International Journal of Climatology 22*, 1715–25.

5   Roy DB & Sparks TH (2000) 'Phenology of British butterflies and climate change.' *Global Change Biology 6*, 407–16.

6   Amano T, Smithers RJ, Sparks TH & Sutherland WJ (2009) 'A 250-year index of first flowering dates and its response to temperature changes.' *Proceedings of the Royal Society B 277*, 2451–7.

7   Parmesan C (1999) 'Effects of climate change on butterfly distributions.' In RE Green, M Harley, M Spalding & C Zöckler (eds) *Impacts of Climate Change on Wildlife*. Royal Society for the Protection of Birds.

8   Fitter A & & Fitter RSR (2002) 'Rapid changes in flowering time in British plants.' *Science 296*, 1689–91.

9   Wikipedia (2012) *Elfstedentocht*, viewed 22 May 2012 at en.wikipedia.org.

10 Kearney MR, Briscoe NJ, Karoly DJ, Porter WP, Norgate M & Sunnucks P (2010) 'Early emergence in a butterfly causally linked to anthropogenic warming.' *Biology Letters 6*, 674–7.

11 Wisniak J (2000) 'The thermometer – From the Feeling to the instrument.' *Chemical Educator, 5*, 88–91.

12 Jones PD & Wigley TML (2010) 'Estimation of global temperature trends: What's important and what isn't.' *Climatic Change 100*, 59–69.

13 Berkeley Earth Surface Temperature. (2011) Viewed 22 May 2012 at berkeleyearth.org.

14 Meehl GA, Tebaldi C, Walton G, Easterling D & McDaniel L (2009) 'The relative increase of record high maximum temperatures compared to record low minimum temperatures in the US.' *Geophysical Research Letters 36*, L23701.

3   WEATHER IS NOT CLIMATE

1   Hoyt D & Schatten KH (1998) 'A new solar activity reconstruction.' *Solar Physics* *179*, 189–219.

2   Zolitschka B (1992) 'Climatic change evidence and lacustrine varves from maar lakes, Germany.' *Climate Dynamics 6*, 229–32.
    Taylor G, Gasse F, Walker PH & Morgan PJ (1990) 'The palaeoecological and palaeoclimatic significance of Miocene freshwater diatomite deposits from southern New South Wales, Australia.' *Palaeogeography, Palaeoclimatology, Palaeoecology 77*, 127–43.

3   Lockwood M (2009) 'Solar change and climate: An update in the light of the current exceptional solar minimum.' *Proceedings of the Royal Society A 466*, 303–29.

4   Scafetta N & West BJ (2008) 'Is climate sensitive to solar variability?' *Physics Today*, March, 50–1.

5   Solanki SK & Krivova NA (2003) 'Can solar variability explain global warming since 1970?' *Journal of Geophysical Research A 108*, 1200.

6   Lean JL (2010) 'Cycles and trends in solar irradiance and climate.' *Wiley Interdisciplinary Reviews: Climate Change 1*, 111–22.

7   Svensmark H, Bondo T & Svensmark J (2009) 'Cosmic ray decreases affect atmospheric aerosols and clouds.' *Geophysical Research Letters 36*, L15101.

8   Calogovic J, Albert C, Arnold F, Beer J, Desorgher L & Flueckiger EO (2010) 'Sudden cosmic ray decreases: No change of global cloud cover.' *Geophysical Research Letters 37*, L03802.

9   Kulmala M, Riipinen I, Nieminen T, Hulkkonen M, Sogacheva L, Manninen HE, Paasonen P, Petäjä T, Dal Maso M, Aalto PP, Viljanen A, Usoskin I, Vainio R, Mirme S, Minikin A, Petzold A, Hõrrak U, Plaß-Dülmer C, Birmili W & Kerminen V-M (2009) 'Atmospheric data over a solar cycle: No connection between galactic cosmic rays and new particle formation.' *Atmospheric Chemistry and Physics Discussions 9*, 21525–60.

10 Pierce JR & Adams PJ (2009) 'Can cosmic rays affect cloud condensation nuclei by altering new particle formation rates?' *Geophysical Research Letters 36*, L09820.

11 Shrestha G, Traina SJ & Swanston CW (2010) 'Black carbon's properties and role in the environment: A comprehensive review.' *Sustainability 2*, 294–320.

12 Ohmura A (2006) 'Observed long term variations of solar irradiance at the Earth's surface.' *Space Science Reviews 125*, 111–28.

13 Philipona R, Behrens K, and Ruckstuhl C (2009) 'How declining aerosols and rising greenhouse gases forced rapid warming in Europe since the 1980s.' *Geophysical Research Letters 36*, L02806.

14 Trenberth KE & Shea DJ (2006) 'Atlantic hurricanes and natural variability in 2005.' *Geophysical Research Letters 33*, L12704.

15 Schlesinger ME & Ramankutty N (2011) 'An oscillation in the global climate system of period 65–70 years.' *Nature 367*, 723–6.

16 Knudsen MF, Seidenkrantz M-S, Jacobsen BH & Kuijpers A (2011) 'Tracking the Atlantic Multidecadal Oscillation through the last 8,000 years.' *Nature Communications 2* (178).

17 Gray ST, Graumlich LJ, Betancourt JL & Pederson GT (2004) 'A tree-ring based reconstruction of the Atlantic Multidecadal Oscillation since 1567 AD.' *Geophysical Research Letters 31*, L12205.

18 Vinczi M & Jánosi IM (2011) 'Is the Atlantic Multidecadal Oscillation (AMO) a statistical phantom?' *Nonlinear Processes in Geophysics 18*, 469–75.

19 Kerr R (2012) 'A North Atlantic climate pacemaker for the centuries.' *Science 288*, 1984–5.

20 Mazzarella A & Scafetta N (2012) 'Evidences for a quasi 60-year North Atlantic Oscillation since 1700 and its meaning for global climate change.' *Theoretical and Applied Climatology. 107*, 599–609.

21 Luterbacher J, Xoplaki E, Dietrich D, Jones PD, Davies TD, Portis D, Gonzalez-Rouco JF, von Storch H, Gyalistras D, Casty C & Wanner H. (2001) 'Extending North Atlantic Oscillation reconstructions back to 1500.' *Atmospheric Science Letters 2*, 114–24.

    Cook ER, D'Arrigo RD & Mann ME (2002) 'A well-verified, multiproxy reconstruction of the winter North Atlantic Oscillation Index since AD 1400.' *Journal of Climate 15*, 1754–64.

22 Minobe S (1997) 'A 50–70 year climatic oscillation over the North Pacific and North America.' *Geophysical Research Letters 24*, 683–6.

23 Linsley BK, Wellington GM & Schrag DP (2000) 'Decadal sea surface temperature variability in the subtropical South Pacific from 1726 to 1997 AD.' *Science 290*, 1145–8.

24 Ummenhofer CC, England MH, McIntosh PC, Meyers GA, Pook MJ, Risbey JS, Sen Gupta A & Taschetto AS (2009) 'What causes southeast Australia's worst droughts?' *Geophysical Research Letters 36*, L04706.

25 Mayewski PA, Meredith MP, Summerhayes CP, Turner J, Worby A, Barrett PJ, Casassa G, Bertler NAN, Bracegirdle T, Naveira Garabato AC, Bromwich D, Campbell H, Hamilton GS, Lyons WB, Maasch KA, Aoki S, Xiao C & van Ommen T (2009) 'State of the Antarctic and Southern Ocean climate system.' *Reviews of Geophysics 47*, RG1003.

26 Lambin EF, Geist HJ, & Lepers E (2003) 'Dynamics of land-use and land-cover change in tropical regions.' *Annual Review of Environment and Resources 28*, 205–41.

27 Turner BL (ed.) (1990) *The Earth as Transformed by Human Action: Global and Regional Changes in the Biosphere over the Past 300 Years*. Cambridge University Press.

28 Gordon LJ, Steffen W, Bror FJ, Folke C, Falkenmark M & Johannessen Å (2005) 'Human modification of global water flows from the land surface.' *Proceedings of the National Academy of Sciences USA 102*, 7612–17.

29 Miller G, Mangan J, Pollard D, Thompson S, Felzer B & Magee, J (2005) 'Sensitivity of the Australian monsoon to insolation and vegetation: Implications for human impact on continental moisture balance.' *Geology 33*, 65–8.

30 Pitman AJ & Hesse PP (2007) 'The significance of large-scale land cover change on the Australian palaeomonsoon.' *Quaternary Science Reviews 26*, 189–200.

31 McAlpine CA, Syktus JI, Ryan JG, Deo RC, McKeon GM, McGowan HA & Phinn SR (2009) 'A continent under stress: Interactions, feedback and risks associated with impact of modified land cover on Australia's climate.' *Global Change Biology 15*, 2206–23.

32 Bonan GB (2008) 'Forests and climate change: Forcings, feedbacks, and the climate benefits of forests.' *Science 320*, 1444–9.

4 THE THERMOSTAT

1   Arrhenius S (1896) 'On the influence of carbonic acid in the air upon the temperature of the ground.' *London, Edinburgh, and Dublin Philosophical Magazine and Journal of Science [Fifth Series] 41,* 237–76.

2   MacFarling-Meure C, Etheridge D, Trudinger C, Steele P, Langenfelds R, van Ommen T, Smith A & Elkins J (2006) 'Law Dome CO_2, CH_4 and N_2O ice core records extended to 2000 years BP.' *Geophysical Research Letters 33,* L14810.

3   Mann ME, Zhang Z, Rutherford S, Bradley RS, Hughes MK, Shindell D, Ammann C, Faluvegi G & Ni F (2009) 'Global signatures and dynamical origins of the Little Ice Age and Medieval Climate Anomaly.' *Science 326,* 1256–60.

4   Keeling CD (1998) 'Rewards and penalties of monitoring the Earth.' *Annual Reviews of Energy and Environment 23,* 25–82.

5   Callendar GS (1938) 'The artificial production of carbon dioxide and its influence on climate.' *Quarterly Journal of the Royal Meteorological Society 64,* 223–40.

6   Chen C, Harries J, Brindley H & Ringer M (2007) 'Spectral signatures of climate change in the Earth's infrared spectrum between 1970 and 2006.' *Proceedings of the 2007 EUMETSAT Meteorological Satellite Conference & 15th AMS Conference on Satellite Meteorology and Oceanography,* Amsterdam, 24–28 September.

7   Kiehl JT & Trenberth KE (1997) 'Earth's annual global mean energy budget.' *Bulletin of the American Meteorological Society 78,* 197–208.

8   Miskolczi FM (2007) 'Greenhouse effect in semi-transparent planetary atmospheres.' *Quarterly Journal of the Hungarian Meteorological Service 111,* 1–40.

9   Trenberth KE, Fasullo J & Smith L (2005) 'Trends and variability in column-integrated atmospheric water vapor.' *Climate Dynamics 24,* 741–58. Morland J, Coen MC, Hocke K, Jeannet P & Mätzler C (2009) 'Tropospheric water vapour above Switzerland over the last 12 years.' *Atmospheric Chemistry and Physics 9,* 5975–88.

10 Hug H (2007) '*Die Klimakatastrophe – ein spektroskopisches Artefakt?*' viewed earlier 24 May 2012 at www.eike-klima-energie.eu/uploads/media/Hug_Spektroskopisches_Artefakt_2007.pdf.

11 Chen C, Harries J, Brindley H & Ringer M (2007) 'Spectral signatures of climate change in the Earth's infrared spectrum between 1970 and 2006.' *Proceedings of the 2007 EUMETSAT Meteorological Satellite Conference & 15th AMS Conference on Satellite Meteorology and Oceanography,* Amsterdam, 24–28 September.

12 Norris JR (2005) 'Multidecadal changes in near-global cloud cover and estimated cloud cover radiative forcing.' *Journal of Geophysical Research 110,* D08206.

13 Dessler AE (2011) 'Cloud variations and the Earth's energy budget.' *Geophysical Research Letters 38,* L19701.

14 Lemonick MD (2010) 'The effect of clouds on climate: A key mystery for researchers.' *Environment 360,* online publication of the Yale School of Forestry and Environment, http://e360.yale.edu.

15 *Climate Change 2007 – The Physical Science Basis.* Contribution of Working Group I to the Fourth Assessment Report of the IPCC, Chapter 3.

16   *Climate Change 2007 – The Physical Science Basis.* Contribution of Working
      Group I to the Fourth Assessment Report of the IPCCIPCC, Chapter 2.

17   Cox P & Jones C (2008) 'Illuminating the modern dance of climate and CO_2.'
      Science 321, 1642–4.

5   DROUGHTS AND FLOODING RAINS

1   *Time Magazine*, 19 December 1969.

2   Prescott JA & Piper CS (1932) 'The soils of the South Australian Mallee.'
      Transactions of the Royal Society of South Australia 56, 118–47.

3   Lavery B, Kariko A & Nicholls N (1992) 'A historical rainfall data set for
      Australia.' *Australian Meteorological Magazine 40*, 33–9.

4   Held IM, Delworth TL, Lu J, Findell KL, & Knutson TR (2005) 'Simulation of
      Sahel drought in the 20th and 21st centuries.' *Proceedings of the National Academy
      of Sciences USA 102*, 17891–6.

5   *The Christian Science Monitor*, 12 February 2010.

6   Vincent LA & Mekis E (2006) 'Changes in daily and extreme temperature and
      precipitation indices for Canada over the twentieth century.' *Atmosphere-Ocean
      44*, 177–93.

7   Ge S, Yang D & Kane DL (2009) 'Yukon River hydrology and climate
      changes, 1977–2006.' *American Geophysical Union, Fall Meeting 2009*, abstract
      #H53C-0938.

8   Schneider U, Fuchs T, Meyer-Christoffer A & Rudolf B (2008) *Global
      Precipitation Analysis Products of the GPCC.* Global Precipitation Climatology
      Centre, Deutscher Wetterdienst, Offenbach a.M., Germany.

9   Chung CE & Ramanathan V (2006) 'Weakening of North Indian SST gradients
      and the monsoon rainfall in India and the Sahel.' *Journal of Climate 19*, 2036–45.

10   Nigam S & Guan B (2011) 'Atlantic tropical cyclones in the twentieth century:
      Natural variability and secular change in cyclone count.' *Climate Dynamics 36*,
      2279–93.

11   Webster PJ, Holland GJ, Curry JA & Chang H-R (2005) 'Changes in tropical
      cyclone number, duration, and intensity in a warming environment.' *Science 309*,
      1844–6.

12   Atlantic Oceanographic and Meteorological Laboratory, Oceanic and
      Atmospheric Research Facility, National Oceanic and Atmospheric
      Administration, US Department of Commerce.

13   Ho C-H, Baik J-J, Kim J-H, Gong D-Y & Sui C-H (2004) 'Interdecadal changes
      in summertime typhoon tracks.' *Journal of Climate 17*, 1767–76.

14   Liu K-b, Shen C & Louie K-s (2001) 'A 1,000-year history of typhoon
      landfalls in Guangdong, Southern China, reconstructed from Chinese historical
      documentary records.' *Annals of the Association of American Geographers 91*, 453–64.

15   Nott J, Haig J, Neil H & Gillieson D (2007) 'Greater frequency variability of
      landfalling tropical cyclones at centennial compared to seasonal and decadal
      scales.' *Earth and Planetary Science Letters 255*, 367–72.

16   From the transcription of his diary by Jan and Frank Nicholas, University of
      Sydney.

17   Gibbs LM (2009) 'Just add water: Colonisation, water governance, and the
      Australian inland.' *Environment and Planning A 41*, 2964–83.

18 Ummenhofer CC, England MH, McIntosh PC, Meyers GA, Pook MJ, Risbey JS, Sen Gupta A & Taschetto AS (2009) 'What causes southeast Australia's worst droughts?' *Geophysical Research Letters 36*, L04706.

19 Piechota T, Timilsena J, Tootle G & Hidalgo H (2004) 'The western US drought: How bad is it?' *Eos 85*, 301.

20 Dai A, Trenberth KE, & Qian T (2004) 'A global dataset of Palmer Drought Severity Index for 1870–2002: Relationship with soil moisture and effects of surface warming.' *Journal of Hydrometeorology 5*, 1117–30.

21 Post DA, Chiew FHS, Teng J, Vaze J, Yang A, Mpelasoka F, Smith I, Katzfey J, Marston F, Marvanek S, Kirono D, Nguyen K, Kent D, Donohue R, Li L & McVicar T (2009) *Production of Climate Scenarios for Tasmania.* A report to the Australian Government from the CSIRO Tasmania Sustainable Yields Project, CSIRO Water for a Healthy Country Flagship, Australia.

22 *Canberra Times*, 19 January 2010.

23 Barriendos M & Rodrigo FS (2006) 'Study of historical flood events on Spanish rivers using documentary data.' *Hydrological Sciences Journal 51*, 765–83.

24 Cyberski J, Grze M, Gutry-Korycka M, Nachlik E & Kundzewicz ZW (2006) 'History of floods on the River Vistula.' *Hydrological Sciences Journal 51*, 799–817.

25 Böhm O & Wetzel K-F (2006) 'Flood history of the Danube tributaries Lech and Isar in the Alpine foreland of Germany.' *Hydrological Sciences Journal 51*, 784–98.

26 Glaser R, Riemann D, Schönbein J, Barriendos M, Brázdil R, Bertolin C, Camuffo D, Deutsch M, Dobrovolný P & van Engelen A (2010) 'The variability of European floods since AD 1500.' *Climatic Change 101*, 235–56.

27 Australian Bureau of Meteorology (2010) *Known floods in the Brisbane and Bremer River Basin*. Online data from www.bom.gov.au.

28 Ralph TJ & Hesse PP (2010) 'Downstream hydrogeomorphic changes along the Macquarie River, southeastern Australia, leading to channel breakdown and floodplain wetlands.' *Geomorphology 118*, 48–64.

Bunn SE, Thoms MC, Hamilton SK, Capon SJ (2006) 'Flow variability in dryland rivers: Boom, bust and the bits in between.' *River Research and Applications 22*, 179–86.

29 Rasmusson EM & Carpenter TH (1982) 'Variations in tropical sea surface temperatures and surface wind fields associated with the Southern Oscillation/El Niño.' *Monthly Weather Review 110*, 354–84.

6 SNOW AND ICE

1 Steiner D, Zumbühl HJ & Bauder A (2008) 'Two Alpine glaciers over the past two centuries.' In B Orlov, E Wiegandt & B Luckman (eds). *Darkening Peaks: Glacier Retreat, Science and Society*. University of California Press.

2 Thompson LG, Mosley-Thompson E, Davis ME, Lin P-N, Henderson K & Mashiotta TA. (2003) 'Tropical glacier and ice core evidence of climate change on annual to millennial time scales.' *Climatic Change 59*, 137–55.

3 Thompson LG, Brecher HH, Mosley-Thompson E, Hardy DR & Mark BG (2009) 'Glacier loss on Kilimanjaro continues unabated.' *Proceedings of the National Academy of Sciences USA 106*, 19770–5.

4   Mölg T, Cullen NJ, Hardy DR, Winkler M & Kaser G (2009) 'Quantifying climate change in the tropical midtroposphere over East Africa from glacier shrinkage on Kilimanjaro.' *Journal of Climate 22*, 4162–81.

5   Gjermundsen E (2007) 'Recent changes in glacier area in the Central Southern Alps of New Zealand.' Master's thesis, University of Oslo.

6   Hoelzle M, Chinn T, Stumm D, Paul F, Zemp M & Haeberli W (2007) 'The application of glacier inventory data for estimating past climate change effects on mountain glaciers: A comparison between the European Alps and the Southern Alps of New Zealand.' *Global and Planetary Change 56*, 69–82.

7   Plummer M (2004) 'Sensitivity of alpine glaciers to climate change and predictions of the demise of Wyoming's glaciers.' *Geological Society of America Abstracts with Programs, 36*(5), 249.

8   Kulkarni AV, Bahuguna IM, Rathore BP, Singh SK, Randhawa SS, Sood RK & Dhar S (2007) 'Glacial retreat in Himalaya using Indian remote sensing satellite data.' *Current Science 92*, 69–74.

9   Bajracharya SR, Mool PK & Shrestha BR (2007) *Impact of Climate Change on Himalayan Glaciers and Glacier Lakes.* International Centre for Integrated Mountain Development.

10   Bishop MP, Bush AB, Collier E, Copland L, Haritashya UK, John SF, Swenson SC & Wahr J (2008) 'Advancing glaciers and positive mass anomaly in the Karakoram Himalaya, Pakistan.' *American Geophysical Union, Fall Meeting 2008, abstract #C32B-04.*

11   Dyurgerov MB & Meier M (2005) 'Glaciers and the changing Earth system: A 2004 snapshot.' Occasional Paper 58. Institute of Arctic and Alpine Research, University of Colorado, Boulder, Colorado.

12   Wiles GC, D'Arrigo RD, Villalba R, Calkin PE & Barclay DJ (2004) 'Century-scale solar variability and Alaskan temperature change over the past millennium.' *Geophysical Research Letters 31*, L15203.

     Reyes AV, Wiles GC, Smith DJ, Barclay DJ, Allen S, Jackson S, Larocque S, Laxton S, Lewis D, Calkin PE & Clague JJ (2006) 'Expansion of alpine glaciers in Pacific North America in the first millennium AD.' *Geology 34*, 57–60.

13   Nesje A, Bakke J, Dahl SO, Øyvind L & Matthews JA (2008) 'Norwegian mountain glaciers in the past, present and future.' *Global and Planetary Change 60*, 10–27.

14   Braithwate RJ & Raper SCB (2002) 'Glaciers and their contribution to sea level change.' *Physics and Chemistry of the Earth 27*, 1445–54.

15   Mayewski PA, Rohling EE, Stager JC, Karlen W, Maasch KA, Meeker LD, Meyerson EA, Gasse F, van Kreveld S, Holmgren K, Lee-Thorp J, Rosqvist G, Rack F, Staubwasser M, Schneider RR & Steig EJ (2004) 'Holocene climate variability.' *Quaternary Research 62*, 243–255.

16   Romanovsky VE, Smith SL & Christiansen HH (2010) 'Permafrost thermal state in the Polar Northern Hemisphere during the International Polar Year 2007–2009: A synthesis.' *Permafrost and Periglacial Processes 21*, 106–16.

17   Masson-Delmotte V, Kageyama M, Braconnot P, Charbit S, Krinner G, Ritz C, Guilyardi E, Jouzel J, Abe-Ouchi A, Crucifix M, Gladstone RM, Hewitt CD, Kitoh A, LeGrande AN, Marti O, Merkel U, Motoi T, Ohgaito R, Otto-Bliesner B, Peltier WR, Ross I, Valdes PJ, Vettoretti G, Weber SL, Wolk F & Yu, Y (2006)

'Past and future polar amplification of climate change: Climate model intercomparisons and ice-core constraints.' *Climate Dynamics 26*, 513–29.

18   Peterson BJ, Holmes RM, McClelland JW, Vörösmarty CJ, Lammers RB, Shiklomanov AI, Shiklomanov IA & Rahmstorf S (2002) 'Increasing river discharge to the Arctic Ocean.' *Science 298*, 2171–3.

19   Kwok R & Rothrock DA (2009) 'Decline in Arctic sea ice thickness from submarine and ICESat records: 1958–2008.' *Geophysical Research Letters 36*, L15501.

20   McKay JL, de Vernal A, Hillaire-Marcel C, Not C, Polyak L & Darby D (2008) 'Holocene fluctuations in Arctic sea-ice cover: Dinocyst-based reconstructions for the eastern Chukchi Sea.' *Canadian Journal of Earth Sciences 45*, 1377–97.

21   Turner J, Comiso JC, Marshall GJ, Lachlan-Cope TA, Bracegirdle T, Maksym T, Meredith MP, Wang Z & Orr A (2009) 'Non-annular atmospheric circulation change induced by stratospheric ozone depletion and its role in the recent increase of Antarctic sea ice extent.' *Geophysical Research Letters 36*, L08502.

22   Rignot E, Bamber JL, van den Broeke MR, Davis C, Li Y, van de Berg WJ, & van Meijgaard EK. (2007) 'Recent Antarctic ice mass loss from radar interferometry and regional climate modeling.' *Nature Geoscience 1*, 106–10.

23   Zwally HJ & Giovinetto MB (2011) 'Overview and assessment of Antarctic ice-sheet mass balance estimates: 1992–2009.' *Surveys in Geophysics 32*, 351–76.

24   Velicogna I (2009) 'Increasing rates of ice mass loss from the Greenland and Antarctic ice sheets revealed by GRACE.' *Geophysical Research Letters 36*, L19503.

25   Chen JL, Wilson CR, Tapley BD, Blankenship D & Young D (2008) 'Antarctic regional ice loss rates from GRACE.' *Earth and Planetary Science Letters 266*, 140–8.

26   Lee H, Shum CK, Howat IM, Monaghan A, Ahn Y, Duan J, Guo J-Y, Kuo C-Y & Wang L (2012) 'Continuously accelerating ice loss over Amundsen Sea catchment, West Antarctica, revealed by integrating altimetry and GRACE data.' *Earth and Planetary Science Letters 321–2*, 74–80.

27   Chen JL, Wilson CR & Tapley BD (2006) 'Satellite gravity measurements confirm accelerated melting of Greenland ice sheet.' *Science 313*, 1958–60.

28   Rignot E, Box JE, Burgess E & Hanna E (2008) 'Mass balance of the Greenland ice sheet from 1958 to 2007.' *Geophysical Research Letters 35*, L20502.

29   Velicogna I (2009) 'Increasing rates of ice mass loss from the Greenland and Antarctic ice sheets revealed by GRACE.' *Geophysical Research Letters 36*, L19503.

30   Johannessen OM, Khvorostovsky K, Miles MW & Bobylev LP (2005) 'Recent ice-sheet growth in the interior of Greenland.' *Science 310*, 1013–16.

31   Rignot E, Velicogna I, van den Broeke MR, Monaghan A & Lenaerts J (2011) 'Acceleration of the contribution of the Greenland and Antarctic ice sheets to sea level rise.' *Geophysical Research Letters 38*, L05503.

32   Boon S, Burgess DO, Koerner RM & Sharp J (2010) 'Forty-seven years of research on the Devon Island Ice Cap, Arctic Canada.' *Arctic 63*, 13–29.

33   Shepherd A, Zhijun DU, Benham TJ, Dowdeswell JA & Morris EM (2007) 'Mass balance of Devon Ice Cap, Canadian Arctic.' *Annals of Glaciology 46*, 249–54.

7   THE OCEAN

1   Frakes LA, Burger D, Apthorpe M, Wiseman J, Dettmann M, Alley N, Flint R, Gravestock D, Ludbrook N, Backhouse J, Skwarko S, Scheibnerova V, McMinn A,

Moore PS, Bolton BR, Douglas JG, Christ R, Wade M, Molnar RE, McGowran B, Balme BE. & Day RA (Australian Cretaceous Palaeoenvironments Group) (1987) 'Australian Cretaceous shorelines, stage by stage.' *Palaeogeography, Palaeoclimatololgy, Palaeoecology 59*, 31–48.

2 O'Connell JF, Allen J & Hawkes K (2010) 'Pleistocene Sahul and the Origins of Seafaring.' In A Anderson, J Barrett & K Boyle (eds) *The Global Origins and Development of Seafaring*. McDonald Institute for Archaeological Research, Cambridge University.

3 Domingues CM, Church JA, White NJ, Gleckler PJ, Wijffels SE, Barker PM & Dunn JR (2008) 'Improved estimates of upper-ocean warming and multi-decadal sea-level rise.' *Nature 453*, 1090–3.

4 Purkey SG & Johnson GC (2010) 'Warming of global abyssal and deep Southern Ocean waters between the 1990s and 2000s: Contributions to global heat and sea level rise budgets.' *Journal of Climate 23,* 6336–51.

5 Wigley TML (2005) 'The climate change commitment.' *Nature 307*, 1766–9.

6 Brohan P, Kennedy JJ, Harris I, Tett SFB & Jones PD (2006) 'Uncertainty estimates in regional and global observed temperature changes: A new dataset from 1850.' *Journal of Geophysical Research 111*, D12106.

7 McLean JD, de Freitas CR & Carter RM (2009) 'Influence of the Southern Oscillation on tropospheric temperature.' *Journal of Geophysical Research 114*, D14104.

8 Foster G, Annan JD, Jones PD, Mann ME, Mullan B, Renwick J, Salinger J, Schmidt GA & Trenberth KE (2009) 'Comment on "Influence of the Southern Oscillation on tropospheric temperature" by JD McLean, CR de Freitas & RM Carter.' *Journal of Geophysical Research Atmospheres 115*, D09110.

9 Douglas BC (1991) 'Global sea level rise.' *Journal of Geophysical Research 96 (C4)*, 6981–92.

10 Cazanave A, Chambers DP, Cipollini P, Fu LL, Hurell JW, Merrifield M, Nerem S, Plag HP, Shum CK & Willis J (2010) 'Sea level rise: Regional and global trends.' In J Hall, DE Harrison & D Stammer (eds) *Proceedings of Ocean Obs'09: Sustained Ocean Observations and Information for Society* (Vol. 1), Venice, Italy, 21–25 September 2009, ESA Publication WPP-306.

11 Church JA, White NJ, Aarup T, Wilson WS, Woodworth PL, Domingues CM, Hunter JR & Lambeck K (2008) 'Understanding global sea levels: Past, present and future.' *Sustainability Science 3*, 9–22.

12 Cazanave A & Llovel W (2010) 'Contemporary sea level rise.' *Annual Review of Marine Science 2*, 145–73.

13 IPCC (2007) *Climate Change 2007: The Physical Science Basis, Chapter 5. Contribution of Working Group I to the Fourth Assessment Report of the Intergovernmental Panel on Climate Change* [Solomon S, Qin D, Manning M, Chen Z, Marquis M, Averyt KB, Tignor MMB & Miller HL Jr. (eds)]. Cambridge University Press, Cambridge, United Kingdom and New York, USA.

14 Dore JE, Lukas R, Sadler DW, Church MJ & Karl DM (2009) 'Physical and biogeochemical modulation of ocean acidification in the central North Pacific.' *Proceedings of the National Academy of Sciences USA 106*, 12235–40.

15 Liu Y, Liu W, Peng Z, Xiao Y, Wei G, Sun W, He J, Liu G & Chou C-L (2009) 'Instability of seawater pH in the South China Sea during the mid-late Holocene: Evidence from boron isotopic composition of corals.' *Geochimica et Cosmochimica Acta 73*, 1264–72.

16   Wei G, McCulloch MT, Mortimer G, Deng W & Xie L (2009) 'Evidence for ocean acidification in the Great Barrier Reef of Australia.' *Geochimica et Cosmochimica Acta 73*, 2332–46.

17   Pelejero C, Calvo E, McCulloch MT, Marshall JF, Gagan MK, Lough JM & Opdyke BN (2005) 'Preindustrial to modern interdecadal variability in coral reef pH.' *Science 309*, 2204–7.

18   Hoegh-Guldberg O, Mumby PJ, Hooten AJ, Steneck RS, Greenfield P, Gomez E, Harvell CD, Sale PF, Edwards AJ, Caldeira K, Knowlton N, Eakin CM, Iglesias-Prieto R, Muthiga N, Bradbury RH, Dubi A & Hatziolos ME (2007) 'Coral reefs under rapid climate change and ocean acidification.' *Science 318*, 1737–42.

19   Kuffner IB, Andersson AJ, Jokiel PL, Rodgers KS & Mackenzie FT (2008) 'Decreased abundance of crustose coralline algae due to ocean acidification.' *Nature Geoscience 1*, 114–17.

20   Fine M & Tchernov D (2007) 'Scleractinian coral species survive and recover from decalcification.' *Science 315*, 1811.

21   Iglesias-Rodriguez MD, Halloran PR, Rickaby REM, Hall IR, Colmenero-Hidalgo E, Gittins JR, Green DRH, Tyrrell T, Gibbs SJ, Dassow P, Rehm E, Armbrust EV & Boessenkool KP (2008) 'Phytoplankton calcification in a high-CO_2 world.' *Science 230*, 336–40.

22   Riebesell U, Zondervan I, Rost B, Tortell PD, Zeebe RE & Morel FFM (2000) 'Reduced calcification of marine plankton in response to increased atmospheric CO_2.' *Nature 407*, 364–7.

23   Riebesell U, Bellerby RGJ, Engel A, Fabry VJ, Hutchins DA, Reusch TBH, Schulz KG & Morel FMM (2008) 'Comment on "Phytoplankton calcification in a high-CO_2 world".' *Science, 322*, 1466b.

24   Iglesias-Rodriguez MD, Buitenhuis ET, Raven JA, Schofield O, Poulton AJ, Gibbs S, Halloran PR & de Baar HJW (2008) 'Response to comment on "Phytoplankton calcification in a high-CO_2 world".' *Science 322*, 1466c.

25   Halloran PR, Hall IR, Colmenero-Hidalgo E & Rickaby REM (2008) 'A multi-species coccolith volume response to an anthropogenically-modified ocean.' *Biogeosciences Discussion 5*, 2923–30.

26   Roberts D, Howard WR, Moy AD, Roberts JL, Trull TW, Bray SG & Hopcroft RR (2011) 'Interannual pteropod variability in sediment traps deployed above and below the aragonite saturation horizon in the sub-Antarctic Southern Ocean.' *Polar Biology 34*, 1739–50.

27   Doney SC, Fabry VJ, Feely RA & Kleypas JA (2009) 'Ocean acidification: The other CO_2 problem.' *Annual Reviews of Marine Science 1,* 169–92.

28   Berkelmans R, De'ath G, Kininmonth S & Skirving WJ (2004) 'A comparison of the 1998 and 2002 coral bleaching events on the Great Barrier Reef: Spatial correlation, patterns and predictions.' *Coral Reefs 23*, 74–83.

29   Great Barrier Reef Marine Park Authority (2006) *Final Bleaching Summary Report 2005/2006*. Climate Change Response Programme, Townsville, Queensland, Australia.

30   Diaz-Pulido G, McCook LJ, Dove S, Berkelmans R, Roff G, Kline DI, Weeks S, Evans RD, Williamson DH & Hoegh-Guldberg O (2009) 'Doom and boom on a resilient reef: Climate change, algal overgrowth and coral recovery.' *PLoS ONE 4*, e5239.

31   Berkelmans R (2002) 'Time-integrated thermal bleaching thresholds of reefs and their variation on the Great Barrier Reef.' *Marine Ecology Progress Series 229*, 73–82.

8   FROM ICE-HOUSE TO GREENHOUSE

1   Kirschvink JL (1992) 'Late Proterozoic low-latitude global glaciation: The snowball Earth.' In JW Schopf & C Klein (eds) *The Proterozoic Biosphere*, pp. 51–52. Cambridge University Press.

2   Hoffman PF & Schrag DP (2002) 'The snowball Earth hypothesis: Testing the limits of global change.' *Terra Nova 14*, 129–55.

3   Bao HM, Lyons JR & Zhou C (2008) 'Triple oxygen isotope evidence for elevated CO_2 levels after a Neoproterozoic glaciation.' *Nature 453*, 504–6.

4   Le Hir G, Donnadieu Y, Goddéris Y, Meyer-Berthaud B, Ramstein G & Blakey RC (2011) 'The climate change caused by the land plant invasion in the Devonian.' *Earth and Planetary Science Letters 310*, 203–12.

5   Beerling D, Berner RA, Mackenzie FT, Harfoot MB & Pyle JA. (2009) 'Methane and the CH_4-related greenhouse effect over the past 400 million years.' *American Journal of Science 309*, 97–113.

6   Zachos J, Pagani M, Sloan L, Thomas E & Billups K (2001) 'Trends, rhythms and aberrations in global climate 65 Ma to present.' *Science 292*, 686–93.

7   Nisbet EG, Jones SM, Maclennan J, Eagles G, Moed J, Warwick N, Bekki S, Braesicke P, Pyle JA & Fowler CRM (2009) 'Kick-starting ancient warming.' *Nature Geoscience 2*, 156–9.

        Dickens GR, O'Neil JR, Rea DK & Owen RM (1995) 'Dissociation of oceanic methane hydrate as a cause of the carbon isotope excursion at the end of the Paleocene.' *Paleoceanography 10*, 965–71.

8   Hönisch B et al. (2012) 'The geological record of ocean acidification.' *Science 335*, 1056–63.

9   Royer DL, Berner RA, Montañez IP, Tabor NJ & Beerling DJ (2004) 'CO_2 as a primary driver of Phanerozoic climate.' *GSA Today 14*, 4–10.

        Park J & Royer DL (2011) 'Geological constraints on the glacial amplification of Phanerozoic climate sensitivity.' *American Journal of Science 311*, 1–26.

10   Shaviv NJ & Veizer J (2003) 'Celestial driver of Phanerozoic climate?' *GSA Today 13*, 4–10.

11   Ruth U, Barnola J-M, Beer J, Bigler M, Blunier T, Castellano E, Fischer H, Fundel F, Huybrechts P, Kaufmann P, Kipfstuhl S, Lambrecht A, Morganti A, Oerter H, Parrenin F, Rybak O, Severi M, Udisti R, Wilhelms F & Wolff E (2007) ' "EDML1": A chronology for the EPICA deep ice core from Dronning Maud Land, Antarctica, over the last 150 000 years.' *Climate of the Past 3*, 475–84.

12   Petit JR, Jouzel J, Raynaud D, Barkov NI, Barnola J-M, Basile I, Bender M, Chappellaz J, Davis M, Delaygue G, Delmotte M, Kotlyakov VM, Legrand M, Lipenkov VY, Lorius C, Pépin L, Ritz C, Saltzman E & Stievenard M (1999) 'Climate and atmospheric history of the past 420,000 years from the Vostok icecore, Antarctica.' *Nature 399*, 429–36.

13   EPICA community members (2004) 'Eight glacial cycles from an Antarctic ice core.' *Nature 429*, 623–8.

14   Motoyama H (2007) 'The second deep ice coring project at Dome Fuji, Antarctica.' *Scientific Drilling 5*, 41–3.

15   Stenni B, Masson-Delmotte V, Selmo E, Oerter H, Meyer H, Röthlisberger R, Jouzel J, Cattani O, Falourd S, Fischer H, Hoffmann G, Iacumin P, Johnsen SJ, Minster B & Udisti R (2010) 'The deuterium excess records of EPICA Dome C and Dronning Maud Land ice cores (East Antarctica).' *Quaternary Science Reviews 29*, 146–59.

16   Sime LC, Wolff EW, Oliver KIC & Tindall JC (2009) 'Evidence for warmer interglacials in East Antarctic ice cores.' *Nature 462*, 342–5.

17   Miller GH, Alley RB, Brigham-Grette J, Fitzpatrick JJ, Polyak L, Serreze MC & White JWC (2010) 'Arctic amplification: Can the past constrain the future?' *Quaternary Science Reviews 29*, 1779–90.

18   Bekryaev RV, Polyakof IV & Alexeev VA (2010) 'Role of polar amplification in long-term surface air temperature variations and modern Arctic warming.' *Journal of Climate 23*, 3888–906.

19   Masson-Delmotte V, Kageyama M, Braconnot P, Charbit S, Krinner G, Ritz C, Guilyardi E, Jouzel J, Abe-Ouchi A, Crucifix M, Gladstone RM, Hewitt CD, Kitoh A, LeGrande AN, Marti O, Merkel U, Motoi T, Ohgaito R, Otto-Bliesner B, Peltier WR, Ross I, Valdes PJ, Vettoretti G, Weber SL, Wolk F & Yu, Y (2006) 'Past and future polar amplification of climate change: Climate model intercomparisons and ice-core constraints.' *Climate Dynamics 26*, 513–29.

     Hansen J, Sato M, Kharecha P, Russell G, Lea DW & Siddal M (2007) 'Climate change and trace gases.' *Philosophical Transactions of the Royal Society A 365*, 1925–54.

20   Petit et al. (2001) 'Vostok ice dore data for 420,000 years.' IGBP PAGES/World Data Center for Paleoclimatology Data Contribution Series #2001–076. NOAA/NGDC Paleoclimatology Program, Boulder CO, USA.

21   Thompson LG, Mosley-Thompson E, Brecher H, Davis M, León B, Les D, Lin P-N, Mashiotta T & Mountain K (2006) 'Abrupt tropical climate change: Past and present.' *Proceedings of the National Academy of Sciences USA 103*, 10536–43.

22   Seppä H, Bjune AE, Telford RJ, Birks HJB & Veski S (2009) 'Last nine-thousand years of temperature variability in Northern Europe.' *Climate of the Past Discussions 5*, 1521–22.

23   Alley RB (2004) 'GISP2 ice core temperature and accumulation data.' IGBP PAGES/World Data Center for Paleoclimatology Data Contribution Series #2004–013. NOAA/NGDC Paleoclimatology Program, Boulder CO, USA.

24   Alley RB (2000) 'The Younger Dryas cold interval as viewed from central Greenland.' *Quaternary Science Reviews 19*, 213–26.

     Blaga CI, Reichart G-J, Lotter AF, Anselmetti F, & Sinninghe Damsté JS (2010) 'Late Glacial to Holocene abrupt temperature changes recorded by Crenarchaeota in Lake Lucerne (Vierwaldstättersee, Switzerland).' *Geophysical Research Abstracts 12*, EGU2010–6531.

     Calvo E, Pelejero C, De Deckker P & Logan, G (2007) 'Antarctic deglacial pattern in a 30 kyr record of sea surface temperature offshore South Australia.' *Geophysical Research Letters 34*, L13707.

     Dansgaard W, White JWC & Johnsen S (1989) 'The abrupt termination of the Younger Dryas climate event.' *Nature 339*, 532–4.

     Gagan MK, Ayliffe LK, Beck JW, Cole JE, Druffel ERM, Dunbar RB & Schrag DP (2000) 'New views of tropical paleoclimates from corals.' *Quaternary Science Reviews 19*, 45–64.

Griffiths ML, Drysdale RN, Vonhof HB, Gagan MK, Zhao J-x, Ayliffe LK, Hantoro WS, Hellstrom JC, Cartwright I, Frisia S & Suwargadi BW (2010) 'Younger Dryas–Holocene temperature and rainfall history of southern Indonesia from $\delta^{18}O$ in speleothem calcite and fluid inclusions.' *Earth and Planetary Science Letters 295*, 30–6.

Rodrigues T, Grimalt JO, Abrantes F, Naughton F & Flores J-A (2010) 'The last glacial-interglacial transition (LGIT) in the western mid-latitudes of the North Atlantic: Abrupt sea surface temperature change and sea level implications.' *Quaternary Science Reviews 29*, 1853–62.

Severinghaus JP, Sowers T, Brook EJ, Alley RB & Bender ML (1998) 'Timing of abrupt climate change at the end of the Younger Dryas interval from thermally fractionated gases in polar ice.' *Nature 391*, 141–6.

Tibby J (2012) 'The Younger Dryas: Relevant in the Australian region?' *Quaternary International 253*, 47–54.

25   Denton GH, Anderson RF, Toggweiler JR, Edwards RL, Schaefer JM & Putnam AE (2010) 'The last glacial termination.' *Science 328*, 1652–56.

26   Shakun JD, Clark PU, He F, Marcott SA, Mix AC, Liu Z, Otto-Bliesner B, Schmittner A & Bard E (2012) 'Global warming preceded by increasing carbon dioxide concentrations during the last deglaciation.' *Nature 484*, 49–54.

27   Delworth TL, Clark PU, Holland CM, John WE, Kuhlbrodt T, Lynch-Steiglitz J, Morrill C, Seager R, Weaver AJ & Zhang R (2008) *Abrupt Climate Change*. A Report by the US Climate Change Science Program and the Subcommittee on Global Change Research. US Geological Survey, Reston VA, 117–62.

Kobashi T, Severinghaus JP, Brook EJ, Barnola J-M & Grachev AM (2011) 'Precise timing and characterization of abrupt climate change 8200 years ago from air trapped in polar ice.' *Quaternary Science Reviews 26*, 1212–22.

28   Upham W (1895) 'The glacial Lake Agassiz.' *USGS Monograph XXV*.

29   Mayewski PA, Rohling EE, Stager JC, Karlen W, Maasch KA, Meeker LD, Meyerson EA, Gasse F, van Kreveld S, Holmgren K, Lee-Thorp J, Rosqvist G, Rack F, Staubwasser M, Schneider RR & Steig EJ (2004) 'Holocene climate variability.' *Quaternary Research 62*, 243–55.

30   Shaviv NJ & Veizer J (2003) 'Celestial driver of Phanerozoic climate?' *GSA Today 13*, 4–10.

31   Monnin E, Indermühle A, Dällenbach A, Flückiger J, Stauffer B, Stocker TF, Raynaud D & Barnola, J-M (2001) 'Atmospheric CO_2 concentrations over the last glacial termination.' *Science 291*(5501), 112–14.

Caillon N, Severinghaus JP, Jouzel J, Barnola J-M, Kang J & Lipenkov VY (2003) 'Timing of atmospheric CO_2 and Antarctic temperature changes across termination III.' *Science 299*(5613), 1728–31.

32   Kennedy M, Mrofka D & von der Borch C (2008) 'Snowball Earth termination by destabilization of equatorial permafrost methane clathrate.' *Nature 453*, 642–5.

9   THE PAST 2000 YEARS

1   Paasche Ø & Bakke J (2010) 'Defining the Little Ice Age.' *Climate of the Past Discussions 6*, 2159–75.

2   Mann ME & Schmidt GA (2003) 'Ground vs. surface air temperature trends: Implications for borehole surface temperature reconstructions.' *Geophysical Research Letters 30*, 1607.

3   Huang S, Pollack HN & Shen P-Y (2008) 'A late Quaternary climate reconstruction based on borehole heatflux data, borehole temperature data, and the instrumental record.' *Geophysical Research Letters 35*, L13703.

4   Pollack HN, Huang S & Smerdon JE (2006) 'Five centuries of climate change in Australia: The view from underground.' *Quaternary Science 21*, 701–6.

5   D'Arrigo R, Wilson R, Liepert B & Cherubini P (2008) 'On the "Divergence Problem" in Northern Forests: A review of the tree-ring evidence and possible causes.' *Global and Planetary Change 60*, 289–305.

6   Lloyd AH & Bunn AG (2007) 'Responses of the circumpolar boreal forest to 20th century climate variability.' *Environmental Research Letters 2*, 045013.

7   D'Arrigo R, Wilson R & Jacoby G (2006) 'On the long-term context for late twentieth century warming.' *Journal of Geophysical Research 111*, D03103.

8   Loso MG, Anderson RS, Anderson SP & Reimer PJ (2006) 'A 1500-year record of temperature and glacial response inferred from varved Iceberg Lake, southcentral Alaska.' *Quaternary Research 66*, 12–24.

9   Thompson LG, Mosley-Thompson E, Brecher H, Davis M, León B, Les D, Lin P-N, Mashiotta T & Mountain K (2006) 'Abrupt tropical climate change: Past and present.' *Proceedings of the National Academy of Sciences USA 103*, 10536–43.

10   Keigwin L (1996) 'The Little Ice Age and Medieval Warm Period in the Sargasso Sea.' *Science 274*, 1504–8.

11   Diz P, Francés G, Pelejero C, Grimalt JO & Vilas F (2002) 'The last 3000 years in the Ría de Vigo (NW Iberian Margin): Climatic and hydrographic signals.' *Holocene 12*, 459–68.

12   Mann ME, Zhihua Zhang Z, Rutherford S, Bradley RS, Hughes MK, Shindell D, Ammann C, Faluvegi G & Ni F (2009) 'Global signatures and dynamical origins of the Little Ice Age and Medieval Climate Anomaly.' *Science 326*, 1256–60.

13   Mann ME, Zhang Z, Hughes MK, Bradley RS, Miller SK, Rutherford S & Ni F (2008) 'Proxy-based reconstructions of hemispheric and global surface temperature variations over the past two millennia.' *Proceedings of the National Academy of Sciences USA, 105*, 13252.

14   Mann ME & Jones PD (2003) 'Global surface temperatures over the past two millennia.' *Geophysical Research Letters 30*(15), 1820.

15   Trouet V, Esper J, Graham NE, Baker A, Scourse JD & Frank DC (2009) 'Persistent positive North Atlantic oscillation mode dominated the Medieval Climate Anomaly.' *Science 324*, 78–80.

16   McIntyre S & McKitrick R (2005) 'Hockey sticks, principal components and spurious significance.' *Geophysical Research Letters 32*, L03710.

17   Huybers P (2005) 'Comment on "Hockey stick principal components and spurious significance".' *Geophysical Research Letters 32*, L20705.

18   Von Storch H & Zorita E (2005) 'Comment on "Hockey stick principal components and spurious significance" by S McIntyre and R McKitrick.' *Geophysical Research Letters 32*, L20701.

19   Juckes MN, Allen MR, Briffa KR, Esper J, Hegerl GC, Moberg A, Osborn TJ & Weber SL (2007) 'Millennial temperature reconstruction intercomparison and evaluation.' *Climate of the Past 3*, 591–609.

20   Wahl ER & Ammann CM (2007) 'Robustness of the Mann, Bradley, Hughes reconstruction of northern hemisphere surface temperatures: Examination of

criticisms based on the nature and processing of proxy climate evidence.' *Climatic Change 85*, 33–69.

21 Von Gunten L, Grosjean M, Rein B, Urrutia R & Appleby P (2009) 'A quantitative high-resolution summer temperature reconstruction based on sedimentary pigments from Laguna Aculeo, central Chile, back to AD 850.' *Holocene 19*, 873–81.

22 Seppä H, Bjune AE, Telford RJ, Birks HJB & Veski S (2009) 'Last nine-thousand years of temperature variability in Northern Europe.' *Climate of the Past Discussions 5*, 1521–52.

23 Frank DC, Esper J, Raible CR, Büntgen U, Trouet V, Stocker B & Joos F (2010) 'Ensemble reconstruction constraints on the global carbon cycle sensitivity to climate.' *Nature 463*, 527–30.

24 Kouwenberg L, Wagner R, Kürschner W & Visscher H (2005) 'Atmospheric CO_2 fluctuations during the last millennium reconstructed by stomatal frequency analysis of *Tsuga heterophylla* needles.' *Geology 33*, 33–6.

25 Berger A & Loutre MF (1997) 'Long-term variations in insolation and their effects on climate, the LLN experiments.' *Surveys in Geophysics 18*, 147–61.

26 Mayewsi PA & Maasch KA (2006) 'Recent warming inconsistent with natural association between temperature and atmospheric circulation over the last 2000 years.' *Climate of the Past Discussions 2*, 327–55.

10 CARBON DIOXIDE AND METHANE

1 Etheridge DM, Steele LP, Langenfelds RL, Francey RJ, Barnola J-M & Morgan VI (1988) 'Natural and anthropogenic changes in atmospheric CO_2 over the last 1000 years from air in Antarctic ice and firn.' *Journal of Geophysical Research 101* (D2), 4115–28.

2 Data from the US Carbon Dioxide Information Analysis Center.

3 Keeling RF, Piper SC, Bollenbacher AF & Walker SJ (2010) 'Monthly atmospheric $^{13}C/^{12}C$ isotopic ratios for 11 SIO stations.' In Carbon Dioxide Information Analysis Center, *Trends: A Compendium of Data on Global Change*. Oak Ridge National Laboratory, US Department of Energy, Oak Ridge, Tenn.

4 *Newsweek*, 31 December, 1990.

5 Carbon Dioxide Information Analysis Center (2010), www.cdiac.ornl.gov.

6 Boden T, Marland G & Andres B (2011) *Global CO_2 Emissions from Fossil Fuel Burning, Cement Manufacture, and Gas Flaring: 1751–2008*. Carbon Dioxide Information Analysis Center, Oak Ridge National Laboratory, Oak Ridge, Tenn.

7 Czelpak G & Junge C (1974) 'Studies of interhemisphere exchange in the troposphere by a diffusion model.' In MN Frenkiel & RE Munn (eds) *Turbulent Diffusion in Environmental Pollution 18B*, 57–72. Academic Press.

8 Manning AC & Keeling RF (2006) 'Global oceanic and land biotic carbon sinks from the Scripps atmospheric oxygen flask sampling network.' *Tellus 58B*, 95–116.

9 West PC, Gibbs HK, Monfreda C, Wagner J, Barford CC, Carpenter SR, & Foley JA (2010) 'Trading carbon for food: Global comparison of carbon stocks vs. crop yields on agricultural land.' *Proceedings of the National Academy of Sciences USA 107*, 19645–8.

10 Peters GP, Marland G, Le Quéré C, Boden T, Canadell JG & Raupach MR (2012) 'Rapid growth in CO_2 emissions after the 2008–2009 global financial crisis.' *Nature Climate Change 2*, 2–4.

11 Mass of atmosphere ($5 \star 10^{15}$ tonnes)$\star 2.4 \star 1.5 / 10^6$ (the 1.5 changes CO_2 ppm by volume to ppm by weight) = $18 \star 10^9$ tonnes.

12 Le Quéré C, Raupach MR, Canadell JG, Marland G, Bopp L, Ciais P, Conway TJ, Doney SC, Feely RA, Foster P, Friedlingstein P, Gurney K, Houghton RA, House JI, Huntingford C, Levy PE, Lomas MR, Majkut J, Metzl N, Ometto JP, Peters GP, Prentice IC, Randerson JT, Running SW, Sarmiento JL, Schuster U, Sitch S, Takahashi T, Viovy N, van der Werf GR, & Woodward FI (2009) 'Trends in the sources and sinks of carbon dioxide.' *Nature Geosciences 2*, 831–6.

13 Blain S et al. (2007) 'Effect of natural iron fertilization on carbon sequestration in the Southern Ocean.' *Nature 446*, 1070–4.

   Bertram C (2009) *Ocean Iron Fertilization in the Context of the Kyoto Protocol and the Post-Kyoto Process*. Kiel Working Paper No. 1523, Kiel Institute for World Economy.

14 Strong A, Chisholm S, Miller C & Cullen J (2009) 'Ocean fertilization: Time to move on.' *Nature 461*, 347–8.

15 Houweling S (1999) 'Global modeling of atmospheric methane sources and sinks.' PhD thesis, University of Utrecht, Netherlands.

16 Shakhova N, Semiletov I, Salyuk A, Yusupov V, Kosmach D & Gustafsson Ö (2010) 'Extensive methane venting to the atmosphere from sediments of the East Siberian Arctic Shelf.' *Science 327*, 1246–50.

17 *Independent,* 13 November, 2011.

18 Buffett B & Archer D (2004) 'Global inventory of methane clathrate: Sensitivity to changes in the deep ocean.' *Earth and Planetary Science Letters 227*, 185–99.

11 DENIAL

1 Carter RM (2010) *Climate: The Counter Consensus.* Stacey International.

2 Plimer I (2011) *How to Get Expelled From School* (page 7). Connor Court.

3 Callendar GS (1938) 'The artificial production of carbon dioxide and its influence on climate.' *Quarterly Journal of the Royal Meteorological Society 64*: 223–40.

   Spencer Weart's history of the science of climate change gives a fascinating insight into its development. Weart S (2003) *The Discovery of Global Warming*. American Institute of Physics Center for History of Physics, available at www.aip.org/history.

4 Plimer IR (2009) *Heaven and Earth. Global Warming: The Missing Science*. Connor Court.

5 Pearson PM & Palmer MR (2000) 'Atmospheric carbon dioxide concentrations over the past 60 million years.' *Nature 406*, 695–9.

6 Klyashtorin LB & Lyubushin AA (2003) 'On the coherence between dynamics of the world fuel consumption and global temperature anomaly.' *Energy & Environment 14*, 733–82.

7 Willis JK, Lyman JM, Johnson GC & Gilson J (2007) 'Correction to "Recent cooling of the upper ocean".' *Geophysical Research Letters 34*, L16601.

8   Polyakov IV, Bekryaev RV, Alekseev GV, Bhatt US, Colony RL, Johnson MA, Maskshtas AP & Walsh D (2003) 'Variability and trends of air temperature and pressure in the maritime Arctic, 1875–2000.' *Journal of Climate 16*, 2067–77.

9   Pryzbylak R (2007) 'Recent air-temperature changes in the Arctic.' *Annals of Glaciology 46*, 316–24.

10   Mann ME, Bradley RS & Hughes MK (1998) 'Global-scale temperature patterns and climate forcing over the past six centuries.' *Nature 392*, 779–87.

11   Mann ME & Jones PD (2003) 'Global surface temperatures over the past millennia.' *Geophysical Research Letters 30*(15), 1820.

12   Segalstad TV (1998) 'Carbon cycle modelling and the residence time of natural and anthropogenic atmospheric CO_2: On the construction of the "Greenhouse Effect Global Warming" dogma.' In R Bate (ed.) *Global Warming: The Continuing Debate*. European Science and Environment Forum, 184–219.

13   Beck E-G (2007) '180 years of atmospheric CO_2 gas analysis by chemical methods.' *Energy & Environment 18*, 259–82.

14   Bray JR (1958) 'An analysis of the possible recent change in atmospheric carbon dioxide concentration.' *Tellus, XI*, 220–30.

15   Keeling RJ (2007) 'Comment on "180 years of atmospheric CO_2 gas analysis by chemical methods" by Ernst-Georg Beck.' *Energy and Environment 18*, 259–82.

16   Pataki DE, Bowling DR & Ehleringer JR (2003) 'Seasonal cycle of carbon dioxide and its isotopic composition in an urban atmosphere: Anthropogenic and biogenic effects.' *Journal of Geophysical Research 108*(D23), 4735.

17   Brown HT & Escombe F (1905) 'On the variations in the amount of carbon dioxide in the air of Kew during the years 1898–1901.' *Proceedings of the Royal Society B, 76*, 118–21.

18   Fonselius S, Koroleff F & Wärme K-E (1956) 'Carbon dioxide variations in the atmosphere.' *Tellus 8*, 176–83.

19   Santer BD, Thorne PW, Haimberger L, Taylor KE, Wigley TML, Lanzante JR, Solomon S, Free M, Gleckler PJ, Jones PD, Karl TR, Klein SA, Mears C, Nychka D, Schmidt GA, Sherwood SC & Wentz FJ (2008) 'Consistency of modelled and observed temperature trends in the tropical troposphere.' *International Journal of Climatology 28*, 1703–22.

12 BET YOUR GRANDCHILDREN'S LIVES ON IT, TOO?

1   Here is a partial list of factors, taken at random from one such model: ocean biology allows the completion of the carbon cycle, provision of di-methyl sulphide (DMS) emissions from phytoplankton, differentiation between diatom and non-diatom plankton, plankton distributions, rates of productivity and emissions of DMS, a tropospheric chemistry scheme, new aerosol species (organic carbon and dust), coupling between the chemistry and sulphate aerosols, and the tropospheric ozone distribution (HadGEM2 model, Hadley Centre technical note 74, 2008).

2   Lockwood M (2008) 'Recent changes in solar outputs and the global mean surface temperature. III. Analysis of contributions to global mean air surface temperature rise.' *Proceedings of the Royal Society A 464*, 1387–1404.

3   Hansen J, Johnson D, Lacis A, Lebedeff S, Lee P, Rind D & Russell G (1981) 'Climate impact of increasing atmospheric carbon dioxide.' *Science 213*, 957–66.

4   Hansen J, Fung I, Lacis A, Rind D, Lebedeff S, Ruedy R & Russell G (1988) 'Global climate changes as forecast by Goddard Institute for Space Studies three-Dimensional model.' *Journal of Geophysical Research 93*, 9341–64.

US Senate Commission on Energy and Natural Resources (1988) *Greenhouse Effect and Global Climate Change.* Government Printing Office, Washington, DC.

5   Hansen J, Sato M, Ruedy R, Lo K, Lea DW, & Medina-Elizade M (2006) 'Global temperature change.' *Proceedings of the National Academy of Sciences USA 103*, 14288–93.

6   Rahmstorf S, Cazenave A, Church JA, Hansen JE, Keeling RF, Parker DE & Somerville RCJ (2007) 'Recent climate observations compared to projections.' *Science 316*, 709.

Pielke, RA Jr (2008) 'Climate predictions and observations.' *Nature Geoscience 1*, 206.

7   Schmittner A, Urban NM, Shakun JD, Mahowald NM, Clark PU Bartlein PJ, Mix AC & Rosell-Melé A (2011) 'Climate sensitivity estimated from temperature reconstructions of the last glacial maximum.' *Science 334*, 1385–8.

8   Pagani M, Liu Z, LaRiviere J & Ravelo AC (2010) 'High Earth-system climate sensitivity determined from Pliocene carbon dioxide concentrations.' *Nature Geoscience 3*, 27–30.

9   Royer DL, Pagani M & Beerling DJ (2011) 'Geologic constraints on Earth system sensitivity to CO_2 during the Cretaceous and early Paleogene.' *Earth System Dynamics Discussions 2*, 211–40.

10   Charney J (1979), cited in Schmittner et al. (2011) *Carbon Dioxide and Climate: A Scientific Assessment.* National Academy of Sciences Press, Washington, DC.

11   Rogelj J, Meinshausen M & Knutti R (2012) 'Global warming under old and new scenarios using IPCC climate sensitivity range estimates.' *Nature Climate Change 2*, 248–53.

12   Köhler P, Bintanja R, Fischer H, Joos F, Knutti R, Lohmann G & Masson-Delmotte V (2009) 'What caused Earth's temperature variations during the last 800,000 years? Data-based evidence on radiative forcing and constraints on climate sensitivity.' *Quaternary Science Reviews 29*, 129–45.

13   Lindzen RS & Choi Y-S (2011) 'On the observational determination of climate sensitivity and its implications.' *Asia-Pacific Journal of Atmospheric Science 47*, 377–90.

14   Hansen J (2009) *Storms of My Grandchildren.* Bloomsbury.

15   Lindzen RS (2007) 'Taking greenhouse warming seriously.' *Energy and Environment 18*, 937–50.

16   Park, J (2009) 'A re-evaluation of the coherence between global-average atmospheric CO_2 and temperatures at interannual time scales.' *Geophysical Research Letters 36*, L22704.

17   Wigley TML (2005) 'The climate change commitment.' *Science 307*, 1766–9.

18   Eby M, Zickfeld K, Montenegro A, Archer D, Meissner KJ & Weaver AJ (2009) 'Lifetime of anthropogenic climate change: Millennial time scales of potential CO_2 and surface temperature perturbations.' *Journal of Climate 22*, 2501–11.

19   Pittock AB (2008) 'Ten reasons why climate change may be more severe than projected.' In MC MacCracken, F Moore & JC Topping Jr (eds) *Sudden and Disruptive Climate Change.* Earthscan.

Lenton TM, Held H, Kriegler E, Hall JW, Lucht W, Rahmstorf S & Schellnhuber HJ (2008) 'Tipping elements in the Earth's climate system.' *Proceedings of the National Academy of Sciences USA 105*, 1786–93.

20 Sherwood SC & Huber M (2011) 'An adaptability limit to climate change due to heat stress.' *Proceedings of the National Academy of Sciences USA 107*, 9552–5.

21 Lobell DB, Sibley A & Ortiz-Monasterio JI (2012) 'Extreme heat effects on wheat senescence in India.' *Nature Climate Change 2*, 186–9.

22 Ainsworth EA (2008) 'Rice production in a changing climate: A meta-analysis of responses to elevated carbon dioxide and elevated ozone concentration.' *Global Change Biology 14*, 1642–50.

23 Leakey ADB, Ainsworth AA, Bernacchi CJ, Rogers A, Long SP & Ort DR (2009) 'Elevated CO_2 effects on plant carbon, nitrogen, and water relations: six important lessons from FACE.' *Journal of Experimental Botany 60*, 2859–76.

24 Ainsworth EA & Long SP (2005) 'What have we learned from 15 years of free-air CO_2 enrichment (FACE)? A meta-analytic review of the responses of photosynthesis, canopy properties and plant production to rising CO_2.' *New Phytologist 165*, 351–72.

25 Ainsworth L & Gillespie K (2010) 'How will all that extra CO_2 affect crops?' *Change and the Heartland 1(1)*, University of Illinois.

26 Terao T, Miura S, Yanagihara T, Hirose T, Nagata K, Tabuchi H, Kim H-Y, Lieffering M, Okada M & Kobayashi K (2005) 'Influence of free-air CO_2 enrichment (FACE) on the eating quality of rice.' *Journal of the Science of Food and Agriculture 85*, 1861–8.

27 Taub DR, Miller B & Allen H (2008) 'Effects of elevated CO_2 on the protein concentration of food crops: A meta-analysis.' *Global Change Biology 14*, 565–75.

28 Högy P & Fangmeier A (2008) 'Effects of elevated atmospheric CO_2 on grain quality of wheat.' *Journal of Cereal Science 48*, 580–91.

29 Wilson PL, Ringrose-Voase A, Jacquier D, Gregory L, Webb M, Wong MTF, Powell B, Brough D, Hill J, Lynch B, Schoknecht N & Griffin T (2009) 'Land and soil resources in northern Australia.' In *Northern Australia Land and Water Science Review*, CSIRO Publishing.

30 Felderhof L & Gillieson D (2006) 'Comparison of fire patterns and fire frequency in two tropical savanna bioregions.' *Austral Ecology 31*, 736–46.

31 Steffen W, Burbridge AA, Hughes L, Kitching R, Lindenmayer D, Musgrave W, Stafford Smith M & Werner PA (2009) *Australia's Biodiversity and Climate Change*. CSIRO Publishing.

32 *2009 Victorian Bushfires Royal Commission, Volume 1 – The Fires and Fire-Related Deaths, Appendix C*. (2010) www.royalcommission.vic.gov.au.

33 Sinervo et al. (2010) 'Erosion of lizard diversity by climate change and altered thermal niches.' *Science 328*, 894–9.

34 Hansen J (2009) *Storms of My Grandchildren* (page 144). Bloomsbury.

35 Rahmstorf S (2010) 'A new view on sea level rise.' *Nature Reports Climate Change* doi:10.1038/climate.2010.29.

36 Church JA, Hunter JR, McInnes KL & White NJ (2006) 'Sea-level rise around the Australian coastline and the changing frequency of extreme sea-level events.' *Australian Meteorological Magazine 55*, 253–60.

37   Great Barrier Reef Foundation (2009) *Valuing the Effects of Coral Reef Bleaching*. Oxford Economics.

38   Great Barrier Reef Foundation (2009) *The Reef and Climate Change*. www.barrierreef.org.

39   Hoegh-Guldberg O, Anthony K, Berkelmans R, Dove S, Fabricus K, Lough J, Marshall P, van Oppen MJH, Negri A & Willis B (2007) 'Vulnerability of reef-building corals on the Great Barrier Reef to climate change.' In JE Johnson & PA Marshall (eds) *Climate Change and the Great Barrier Reef*. Great Barrier Reef Marine Park Authority and Australian Greenhouse Office.

40   Doney SC, Fabry VJ, Feely RA & Kleypas JA (2009) 'Ocean acidification: The other CO_2 problem.' *Annual Reviews of Marine Science 1*, 169–92.

41   Schellnhuber HJ (2010) 'Tragic triumph.' *Climatic Change 100*, 229–38.

42   *Nuclear Energy Outlook 2008*. OECD 2008.

43   Jacobsen MZ & Delucchi A (2011) 'Providing all global energy with wind, water, and solar power, Part I: Technologies, energy resources, quantities and areas of infrastructure, and materials.' *Energy Policy 39*, 1154–69.

44   Miller LM, Gans F & Kleidon A (2010) 'Estimating maximum global land surface wind power extractability and associated climatic consequences.' *Earth System Dynamics Discussions 1*, 169–89.

45   Broecker WS & Kunzig R (2008) *Fixing Climate: What Past Climate Changes Reveal About the Current Threat – And How to Counter It*. Hill & Wang.

46   Woolf D, Amonette JE, Street-Perrot FA, Lehmann J & Joseph S (2010) 'Sustainable biochar to mitigate global climate change.' *Nature Communications 1*, 56.

47   Crutzen P (2006) 'Albedo enhancement by stratospheric sulfur injections: A contribution to resolve a policy dilemma.' *Climatic Change 77*, 211–20.

48   Smith SJ, van Aardenne J, Klimont Z, Andres RJ, Volke A & Delgado Arias S (2011) 'Anthropogenic sulfur dioxide emissions: 1850–2005.' *Atmospheric Chemistry and Physics 11*, 1101–6.

49   Wood G (2009) 'Re-engineering the Earth.' *Atlantic*, July–August 2009, www.theatlantic.com/magazine

        Hamilton C (2010). 'An evil atmosphere is forming around geoengineering.' *New Scientist 2769*, 22.

50   Wigley TML (2006) 'A combined mitigation/geoengineering approach to climate stabilization.' *Science 314*, 452–4.

51   Caldeira K & Wood L (2008) 'Global and Arctic climate engineering: Numerical model studies.' *Philosophical Transactions of the Royal Society A, 366*, 4039–56.

52   Kerr RA (2006) 'Pollute the planet for climate's sake?' *Science, 314*, 401–3.

53   Ricke KL, Morgan MG & Allen MR (2010) 'Regional climate response to solar-radiation management.' *Nature Geoscience 3*, 537–41.

54   Wigley TML (2011) 'Coal to gas: The influence of methane leakage.' *Climatic Change 108*, 601–8.

INDEX